SpringerBriefs in Optimization

Series Editors

Panos M. Pardalos
János D. Pintér
Stephen M. Robinson
Tamás Terlaky
My T. Thai

SpringerBriefs in Optimization showcases algorithmic and theoretical techniques, case studies, and applications within the broad-based field of optimization. Manuscripts related to the ever-growing applications of optimization in applied mathematics, engineering, medicine, economics, and other applied sciences are encouraged.

For further volumes:
http://www.springer.com/series/8918

Miguel A. Goberna • Marco A. López

Post-Optimal Analysis in Linear Semi-Infinite Optimization

 Springer

Miguel A. Goberna
Statistics and Operations Research
University of Alicante
Alicante, Spain

Marco A. López
Statistics and Operations Research
University of Alicante
Alicante, Spain

ISSN 2190-8354
ISBN 978-1-4899-8043-4
DOI 10.1007/978-1-4899-8044-1
Springer New York Heidelberg Dordrecht London

ISSN 2191-575X (electronic)
ISBN 978-1-4899-8044-1 (eBook)

Library of Congress Control Number: 2013957140

Mathematics Subject Classification (2010): 90C05, 90C34, 90C31

Printed on acid-free paper

Springer is part of Springer Science+Business Media (www.springer.com)

To our wives, Juli and María Pilar
To our collaborators

Preface

Linear semi-infinite optimization (LSIO) deals with linear optimization problems in which either the dimension of the decision space or the number of constraints (but not both) is infinite. A typical feature of this type of optimization problems is that boundedness (i.e., finiteness of the optimal value) does not imply solvability (i.e., existence of an optimal solution). In most LSIO applications, the data defining the nominal problem are uncertain, so that the user must choose among different uncertainty models, e.g., robust models, parametric models, probabilistic models, or fuzzy models, by taking into consideration the nature of the data, the computational effort required to solve the auxiliary problems, the available hardware and software, etc. Parametric models are based on embedding the nominal problem into a suitable topological space of admissible perturbed problems, the so-called space of parameters. Sensitivity analysis provides estimations of the impact of a given perturbation of the nominal problem on the optimal value. Qualitative stability analysis provides conditions under which sufficiently small perturbations of the nominal problem provoke only small changes in the optimal value, the optimal set and the feasible set. Quantitative stability analysis, in turn, yields exact and approximate distances, in the space of parameters, from the nominal problem to important families of problems (e.g., from a bounded problem to the solvable ones) and error bounds (of Lipschitz-type) which are related to the complexity analysis of the numerical methods.

This Springer Brief on post-optimal analysis in LSIO allows us to answer "what if" questions on the basis of stability and sensitivity results whose proofs are generally omitted while their use is illustrated by means of comments and suitable examples. It is intended as a guide for further readings addressed to graduate and postgraduate students of mathematics interested in optimization and also to researchers specialized in parametric optimization and related topics (e.g., algorithmic complexity). Moreover, it could be a useful tool for researchers working in those fields where LSIO models arise in a natural way in uncertain environments (e.g., engineering and finance).

The book is organized as follows. Chapter 1 recalls the necessary preliminaries on the theory and methods of LSIO which are presented in a detailed way in our monograph Linear Semi-Infinite Optimization [102], published in 1998, aggregating some concepts related to complementary solutions which are used in sensitivity analysis and updating the brief review of numerical methods. In Chap. 2 we discuss the advantages and disadvantages of five different approaches to uncertain LSIO which are illustrated by means of the portfolio problem with uncertain returns. The remaining chapters describe the state of the art in those models which have a substantial presence in the LSIO literature: on the robust approach to Linear Semi-Infinite Optimization (Chap. 3), sensitivity analysis (Chap. 4), qualitative stability analysis (Chap. 5), and quantitative stability analysis (Chap. 6). The material reviewed in Chaps. 3, 4, and 6 has been published after 1998 while part of the content of Chap. 5 was already analyzed in detail in [102, Chaps. 6 and 10]. After the introductory Chaps. 1 and 2, Chaps. 3–5 can be read independently, while Chap. 5 contains the preliminaries of Chap. 6. The remarks at the end of each section review the antecedents and extensions of the exposed results and methods, while the last remark of each chapter describes some open problems.

The authors want to thank the coauthors of the many joint works mentioned in this book: J. Amaya, E. Anderson, A. Auslender, P. Bosch, M.J. Cánovas, A. Daniilidis, N. Dinh, A. Dontchev, A. Ferrer, V.E. Gayá, S. Gómez, F. Guerra, A. Hantoute, V. Jeyakumar, V. Jornet, D. Klatte, A. Kruger, M. Larriqueta, G.Y. Li, R. Lucchetti, J.E. Martínez-Legaz, J.A. Mira, B. Mordukhovich, J. Parra, M.M.L. Rodríguez, G. Still, T.Q. Son, T. Terlaky, M. Théra, M.I. Todorov, F.J. Toledo, G. Torregrosa, V.N. Vera de Serio, J. Vicente-Pérez, M. Volle, and C. Zălinescu. From all of them we have learnt much. Our special acknowledgment also to M.J. Cánovas, J. Parra, M.M.L. Rodríguez, M. Théra, F.J. Toledo, and E. Vercher for their support, careful reading of the manuscript, and suggestions for improvement, to our students of the Degree of Mathematics in Alicante A. Navarro and R. Campoy for having drawn some figures, and to the participants in a doctoral course based on the draft taught by one of the authors at Universidad Nacional de San Luis (Argentina), April 2013, whose comments and criticisms helped us to improve the quality of the manuscript.

Alicante, Spain   Miguel A. Goberna
October 2013   Marco A. López

Contents

Chapter 1
Preliminaries on Linear Semi-infinite Optimization

1.1 Optimality and Uniqueness

Ordinary (or finite) linear optimization, linear infinite optimization, and linear semi-infinite optimization (LO, LIO, and LSIO in short) deal with linear optimization problems, where the dimension of the decision space and the number of constraints are both finite, both infinite, and exactly one of them finite, respectively. With few exceptions (as some classical applications collected in [102, Chap. 2], among them a model developed by G. Dantzig in his Ph.D. Thesis on statistical inference, started in 1936 and interrupted by World War II, or some recent work, as [190]), most LSIO problems arising in practice have finitely many decision variables, so that they can be expressed as

$$P : \inf_{x \in \mathbb{R}^n} \langle c, x \rangle$$
$$\text{s.t.} \quad \langle a(t), x \rangle \geq b(t), \ t \in T, \tag{1.1}$$

where $\langle \cdot, \cdot \rangle$ denotes the Euclidean scalar product in $\mathbb{R}^n$, T is an infinite set and the data form the triplet $(c, a, b) \in \mathbb{R}^n \times (\mathbb{R}^n)^T \times \mathbb{R}^T$. The LSIO problem P is said to be *continuous* whenever T is a compact Hausdorff topological space, $b \in \mathcal{C}(T)$ (the linear space of real-valued continuous functions on T), and $a = (a_1(\cdot), \ldots, a_n(\cdot)) \in \mathcal{C}(T)^n$. We also write the problem P in (1.1) in matrix form as follows:

$$P : \inf_{x \in \mathbb{R}^n} c'x$$
$$\text{s.t.} \quad a_t'x \geq b_t, \ t \in T, \tag{1.2}$$

where c' denotes the transpose of the column vector $c \in \mathbb{R}^n$. We denote by $F = \{x \in \mathbb{R}^n : a_t'x \geq b_t, \ t \in T\}$ the *feasible set* of P. Obviously, F is a closed convex set (and, conversely, any closed convex set is the solution set of some linear system as a consequence of the separation theorem).

M.A. Goberna and M.A. López, *Post-Optimal Analysis in Linear Semi-Infinite Optimization*, SpringerBriefs in Optimization, DOI 10.1007/978-1-4899-8044-1_1,

Introducing the so-called *marginal function* $g : \mathbb{R}^n \to \overline{\mathbb{R}} = \mathbb{R} \cup \{+\infty\}$ by

$$g(x) := \sup_{t \in T} (b_t - a_t' x), \qquad (1.3)$$

$F = \{x \in \mathbb{R}^n : g(x) \le 0\}$ and P becomes an ordinary optimization problem with linear objective and a unique constraint:

$$P : \inf_{x \in \mathbb{R}^n} c'x$$
$$\text{s.t.} \qquad g(x) \le 0.$$

The *graph*, the *epigraph*, and the *hypograph* of g are

$$\text{gph } g := \{(x, g(x)) : g(x) \in \mathbb{R}\},$$
$$\text{epi } g := \{(x, \lambda) \in \mathbb{R}^{n+1} : g(x) \le \lambda\},$$

and

$$\text{hypo } g := \{(x, \lambda) \in \mathbb{R}^{n+1} : g(x) \ge \lambda\},$$

respectively. When $F \ne \emptyset$, g is a *proper convex* function. Then, the *domain* of g is the nonempty convex set

$$\text{dom } g := \{x \in \mathbb{R}^n : g(x) < +\infty\}.$$

The convexity of g entails the convexity of the *sublevel sets* of g, $\{x \in \mathbb{R}^n : g(x) \le \lambda\}$, $\lambda \in \mathbb{R}$, i.e., the *quasiconvexity* of g.

Unfortunately, the reformulation of P as an ordinary convex optimization problem is only useful when P is continuous for two reasons which are related to the theoretical analysis and the numerical treatment of P :

1. It is difficult to get geometric information on the feasible set F from g.
2. The computation of the *convex subdifferential* of g at a point $x \in F$,

$$\partial g(x) := \{u \in \mathbb{R}^n : g(y) \ge g(x) + u'(y - x) \ \forall y \in \mathbb{R}^n\},$$

is a hard work (see [128] and references therein). Notice that, when P is continuous, Valadier's formula yields

$$\partial g(x) := \text{conv} \{-a_t : g(x) = b_t - a_t' x, \ t \in T\}.$$

Throughout this chapter P denotes a given LSIO problem with fixed data. We denote by $v(P) := \inf_{x \in F} c'x$ (with the convention that $\inf \emptyset = +\infty$) and $S := \{x \in F : c'x = v(P)\}$ the *optimal value* and the *optimal set* of P, respectively. Obviously, F is a closed convex set and S is an exposed face of the feasible set F, but (in contrast with LO) we may have $S = \emptyset$ even though $v(P) \in \mathbb{R}$.

Let us introduce some basic notation. Following Kortanek [161], we denote by $\mathbb{R}^{(T)}$ the linear space of *generalized finite sequences*, whose elements are the functions $\lambda \in \mathbb{R}^T$ that vanish everywhere on T except on a finite subset of T. The notation $\mathbb{R}^{(T)}$ for the space of dual variables has been standard in semi-infinite programming for several decades and was exported to infinite programming in [71]. Let $\mathbb{R}_+$ and $\mathbb{R}_{++}$ ($\mathbb{R}_-$ and $\mathbb{R}_{--}$) be the sets of nonnegative and positive (nonpositive and negative, respectively) real numbers. The positive cone in $\mathbb{R}^{(T)}$ is denoted by $\mathbb{R}_+^{(T)}$ and the null element of $\mathbb{R}^T$ by 0_T.

Let us introduce first the basic notation on sets used in this book. By

$$\operatorname{conv} X := \left\{ \sum_{t \in T} \lambda_t x_t : x_t \in X \ \forall t \in T, \ \lambda \in \mathbb{R}_+^{(T)}, \text{ and } \sum_{t \in T} \lambda_t = 1 \right\},$$

$$\operatorname{cone} X := \left\{ \sum_{t \in T} \lambda_t x_t : x_t \in X \ \forall t \in T, \text{ and } \lambda \in \mathbb{R}_+^{(T)} \right\},$$

$$\operatorname{aff} X := \left\{ \sum_{t \in T} \lambda_t x_t : x_t \in X \ \forall t \in T, \ \lambda \in \mathbb{R}^{(T)}, \text{ and } \sum_{t \in T} \lambda_t = 1 \right\},$$

and

$$\operatorname{span} X := \left\{ \sum_{t \in T} \lambda_t x_t : x_t \in X \ \forall t \in T, \text{ and } \lambda \in \mathbb{R}^{(T)} \right\},$$

we denote the convex hull, the convex conical hull (with the origin), the affine hull, and the linear hull of a nonempty subset X of a linear space and by int X, bd X, and cl X the interior, the boundary, and the closure of a subset X of a topological space. By definition, all the algebraic hulls are empty whenever $X = \emptyset$, except the convex conical hull, which is the singleton set formed by the zero vector. We denote by $\|\cdot\|_2$, $\|\cdot\|_1$, and $\|\cdot\|_\infty$ the Euclidean, the ℓ_1, and the supremum norm in $\mathbb{R}^n$, with closed unit balls $\mathbb{B}_2$, $\mathbb{B}_1$, and $\mathbb{B}_\infty$, and associated distances $d_2(\cdot,\cdot)$, $d_1(\cdot,\cdot)$, and $d_\infty(\cdot,\cdot)$, respectively. The notation $\|\cdot\|$ is used for a general norm, whereas $d(\cdot,\cdot)$ denotes its associated distance. The zero vector of $\mathbb{R}^n$ is denoted by 0_n.

When $\emptyset \neq X \subset \mathbb{R}^n$, equipped with the topology induced by the Euclidean norm, rint X and rbd X represent the relative interior and the relative boundary of X (i.e., the interior and the boundary of X w.r.t. the topology induced by $\mathbb{R}^n$ in aff X). If X is a closed convex set, the *recession cone* of X is $0^+X := \{v \in \mathbb{R}^n : c + v \in X \ \forall c \in X\}$, and it coincides with the set of all the limits of the form $\lim_{r \to \infty} \mu_r x_r$, where $\mu_r \in \mathbb{R}_+$, $x_r \in X$, $r = 1, 2, \ldots$, and $\mu_r \downarrow 0$. Moreover, dim $X := \dim \operatorname{aff} X$ represents the *dimension* of a convex set X, $X^\circ := \{y \in \mathbb{R}^n : x'y \geq 0 \ \forall x \in X\}$ is the *positive polar* of a convex cone X, and $X^\perp := X^\circ \cap (-X^\circ)$ is the *orthogonal subspace* to a linear subspace X. The *linearity* lin X of a convex cone X is the greatest linear subspace contained in X and $X \cap (\operatorname{lin} X)^\perp$ is the *pointed cone* of X (for which 0_n is an extreme point).

We associate with P, or with its *constraint system* $\sigma = \{a'_t x \geq b_t, t \in T\}$, the following sets:

- The convex hull of *constraints data*:

$$C := \text{conv}\{(a_t, b_t), \ t \in T\} \subset \mathbb{R}^{n+1}.$$

- The *first moment cone:*

$$M := \text{cone}\{a_t, \ t \in T\} \subset \mathbb{R}^n.$$

- The *second moment cone;*

$$N := \text{cone}\{(a_t, b_t), \ t \in T\} = \mathbb{R}_+ C \subset \mathbb{R}^{n+1}.$$

- The *characteristic cone:*

$$K := N + \text{cone}\{(0_n, -1)\} \subset \mathbb{R}^{n+1}.$$

Obviously, $M = \text{Proj}_{\mathbb{R}^n}(N) = \text{Proj}_{\mathbb{R}^n}(K)$, where $\text{Proj}_{\mathbb{R}^n} : \mathbb{R}^n \times \mathbb{R} \rightarrow \mathbb{R}^n$ denotes the *projection mapping* on $\mathbb{R}^n$, i.e., $\text{Proj}_{\mathbb{R}^n}(x, x_{n+1}) = x$. We say that σ is *inconsistent* whenever $F = \emptyset$ and it is *strongly inconsistent* in the particular case that some finite subsystem of σ is inconsistent. The characteristic cone K and its closure cl K (both expressed in terms of the data) capture all relevant information on σ and F, respectively. Concerning the moment cones, N describes the consistency of σ while M describes the boundedness of F.

Theorem 1.1.1 (Existence). *A system σ is inconsistent if and only if $(0_n, 1) \in$ cl K, and it is strongly inconsistent if and only if $(0_n, 1) \in K$.*

Theorem 1.1.1 remains true replacing K with N. It is not the case for the LSIO version of the famous Farkas Lemma. Recall that an inequality $w'x \geq \gamma$ is said to be *consequence* of a consistent system σ whenever $w'x \geq \gamma$ for all $x \in F$.

Theorem 1.1.2 (Non-homogeneous Farkas Lemma). *Let σ be a consistent linear system. Then, a linear inequality $w'x \geq \gamma$ is consequence of σ if and only if $(w, \gamma) \in$ cl K.*

Two consistent systems $\sigma = \{a'_t x \geq b_t, t \in T\}$ and $\tilde{\sigma} = \{\tilde{a}'_t x \geq \tilde{b}_t, t \in T\}$ are *equivalent* if they have the same set of solutions, i.e., if $F = \tilde{F}$; in other words, if they constitute two alternative representations of the same closed convex set F. According to Theorem 1.1.2, we have [102, Theorem 5.10]:

- σ and $\tilde{\sigma}$ are equivalent if and only if cl $K = $ cl $\tilde{K}$.

Moreover, as shown in [102, Chaps. 5 and 9], $F \neq \emptyset$ is:

- bounded $\Leftrightarrow (0_n, -1) \in$ int cl $K = $ int $K \Leftrightarrow M = \mathbb{R}^n$.
- a polyhedral convex set $\Leftrightarrow$ cl K is polyhedral.

- an affine manifold $\Leftrightarrow$ the pointed cone of cl K is cone $\{(0_n, -1)\}$.
- full dimensional $\Leftrightarrow$ cl K is pointed.

Denote $s(x,t) := \langle a(t), x \rangle - b(t)$, $t \in T$. The *slack function* at $x \in \mathbb{R}^n$, $s(x, \cdot)$, allows us to check its feasibility: $x \in F$ if and only if $s(x, \cdot)$ is nonnegative on T. The set of zeros of the slack function at x is the so-called *set of active constraints* (also called *binding constraints*) at $x \in \mathbb{R}^n$:

$$T(x) := \{t \in T : s(x,t) = 0\}.$$

So, computing $T(x)$ leads us to the problem of finding the optimal set of the unconstrained global minimization problem $\inf_{t \in T} s(x,t)$. If the slack function is non-identically zero and all the coefficients in the constraint system σ are analytic functions of the index t, then $T(x)$ is a finite set. When these coefficients are polynomial, computing $T(x)$ consists of solving an algebraic equation.

The *cone of feasible directions* and the *active cone* at $x \in F$ are

$$D(F;x) := \{d \in \mathbb{R}^n : \exists \theta > 0, \ x + \theta d \in F\}$$

and

$$A(x) := \text{cone}\{a_t, \ t \in T(x)\},$$

respectively. If $t \in T(x)$ and $d \in \mathbb{R}^n$ satisfies $x + \theta d \in F$ for some $\theta > 0$, then $b_t + a_t' \theta d = a_t'(x + \theta d) \geq b_t$, so that $a_t'd \geq 0$. So, $D(F;x) \subset A(x)^\circ$ and since cl $A(x) = A(x)^{\circ\circ}$ (by the Farkas Lemma for cones), one has

$$A(x) \subset \text{cl } A(x) \subset D(F;x)^\circ. \tag{1.4}$$

We now introduce four constraint qualifications (one local and three global) which are useful in different frameworks (optimality, duality, stability). We say that P (or σ) satisfies:

- the *local Farkas–Minkowsky constraint qualification* (LFMCQ in brief) at $x \in F$ if every consequence of σ binding at x is consequence of a finite subsystem of σ or, equivalently (by Theorem 1.1.2), if $D(F;x)^\circ \subset A(x)$, which itself is equivalent by (1.4) to $D(F;x)^\circ = A(x)$;
- the *Farkas–Minkowsky constraint qualification* (FMCQ) if every consequence of σ is consequence of a finite subsystem or, equivalently (by Theorem 1.1.2), if K is closed;
- the *Slater constraint qualification* (SCQ) if there exists $\hat{x} \in \mathbb{R}^n$ (called *Slater point*) such that $a_t'\hat{x} > b_t$ for all $t \in T$ or, equivalently, if $\hat{x} \in F$ and $T(\hat{x}) = \emptyset$;
- the *strong Slater constraint qualification* (SSCQ) if there exists $\hat{x} \in \mathbb{R}^n$ (called *SS point*) and $\varepsilon > 0$ such that $a_t'\hat{x} \geq b_t + \varepsilon$ for all $t \in T$ or, equivalently, if $v(P_{SS}) > 0$, where

$$P_{SS} : \sup_{(x,y)\in\mathbb{R}^{n+1}} \quad y$$
$$\text{s.t.} \qquad\qquad a_t'x \geq b_t + y, \ t \in T.$$

Observe that the LSIO problem P_{SS} is continuous if and only if P is continuous, and that checking the condition $v(P_{SS}) > 0$ does not require to solve P_{SS} until optimality. Notice also that replacing each constraint $a_t'x \geq b_t$ in σ with the infinitely many constraints $ka_t'x \geq kb_t - 1$, $k \in \mathbb{N}$, one gets another linear representation of F such that all points of F are SS with $\varepsilon = 1$ (even the points of bd F). The existence of linear systems such that every constraint is inactive at any feasible solution is an inconvenient of LSIO in comparison with LO (feasible direction methods use $T(x_k)$ at the current iterate x_k). This undesirable situation is not possible whenever σ is continuous and does not contain the *trivial inequality* $0_n'x \geq 0$, in which case int F is the set of Slater points [102, Corollary 5.9.1]. Obviously,

$$\text{SSCQ} \Rightarrow \text{SCQ}.$$

Moreover, if P continuous and satisfies SCQ, then the compact convex set

$$\text{conv}\,(C \cup \{(0_n,-1)\}) = \text{conv}\,\{(a_t,b_t)\,, t \in T; (0_n,-1)\}$$

does not contain 0_{n+1} and so $K = \mathbb{R}_+ \text{conv}\,(C \cup \{(0_n,-1)\})$ is closed. Therefore,

$$\left.\begin{array}{r} P \text{ continuous} \\ \text{SCQ holds} \end{array}\right\} \Rightarrow \text{FMCQ holds} \Rightarrow \text{LFMCQ holds at any } x \in F.$$

If $\overline{x} \in F$ satisfies the *Karush–Kuhn–Tucker* (KKT) *condition*

$$c \in A(\overline{x}), \tag{1.5}$$

then $c \in D(F;\overline{x})^\circ$ by (1.4), so that $c'(x - \overline{x}) \geq 0$ for all $x \in F$, i.e., $\overline{x} \in S$. Actually, $c \in D(F;\overline{x})^\circ \Leftrightarrow \overline{x} \in S$.

We now assume the existence of $\overline{\lambda} \in \mathbb{R}_+^{(T)}$ such that $\left(\overline{x},\overline{\lambda}\right) \in \mathbb{R}^n \times \mathbb{R}_+^{(T)}$ is a *saddle point* of the *Lagrange function* of P,

$$L(x,\lambda) := c'x + \sum_{t\in T}\lambda_t(b_t - a_t'x),$$

i.e., we assume that

$$L(\overline{x},\lambda) \leq L\left(\overline{x},\overline{\lambda}\right) \leq L\left(x,\overline{\lambda}\right) \text{ for all } (x,\lambda) \in \mathbb{R}^n \times \mathbb{R}_+^{(T)}. \tag{1.6}$$

Given any $s \in T$, the first inequality in (1.6), with $\lambda_t = \overline{\lambda}_t$ for all $t \neq s \in T$ and $\lambda_s = \overline{\lambda}_s + 1$, yields $b_s - a_s'\overline{x} \leq 0$, so that $\overline{x} \in F$. Moreover, given $x \in F$, we have

$$c'\overline{x} = L\left(\overline{x}, 0_T\right) \leq L\left(\overline{x}, \overline{\lambda}\right) \leq L\left(x, \overline{\lambda}\right) = c'x + \sum_{t \in T} \overline{\lambda}_t (b_t - a_t'x) \leq c'x,$$

so that (1.6) also implies $\overline{x} \in S$.

Observe that the first inequality in (1.6) entails $\sum_{t \in T} \overline{\lambda}_t (b_t - a_t'\overline{x}) = L\left(\overline{x}, \overline{\lambda}\right) - L\left(\overline{x}, 0_T\right) \geq 0$, which together with $\overline{x} \in F$ yields the *complementarity condition*

$$\overline{\lambda}_t (b_t - a_t'\overline{x}) = 0 \text{ for all } t \in T. \tag{1.7}$$

Consider a LSIO problem P such that $\overline{x} \in S \neq F$, which entails $c \neq 0_n$. Replacing each constraint $a_t'x \geq b_t$ with $ka_t'x \geq kb_t - 1$, $k \in \mathbb{N}$, we get another LSIO problem with the same feasible set F and cost vector c, so that the optimal set is still S and any feasible point is SS. For the sake of simplicity we assume that this is the case for the initial problem P. Then, (1.5) $\Rightarrow c = 0_n$ (contradiction) while (1.6) $\Rightarrow$ (1.7) $\Rightarrow \overline{\lambda} = 0_T \Rightarrow L\left(\cdot, \overline{\lambda}\right) = \langle c, \cdot \rangle$ and, from the second inequality in (1.6), $c'\overline{x} \leq c'x$ for all $x \in \mathbb{R}^n$, i.e., $c \in (\mathbb{R}^n)^\circ = \{0_n\}$, which also implies $c = 0_n$. Hence, the conditions (1.5) and (1.6) are sufficient, but not necessary, for the optimality of $\overline{x}$ unless certain CQ holds.

Theorem 1.1.3 (Optimality). *If the LFMCQ holds at $\overline{x} \in F$, then the following statements are equivalent to each other:*

  (i) $\overline{x} \in S$.
  (ii) $c \in A\left(\overline{x}\right)$.
  (iii) *There exists $\overline{\lambda} \in \mathbb{R}_+^{(T)}$ such that $\left(\overline{x}, \overline{\lambda}\right)$ is a saddle point of L.*

If the characteristic cone K captures the relevant information on the feasible set F of a consistent problem, a similar role plays the first moment cone regarding the optimal set S. Indeed, S is bounded if and only if $c \in \text{int } M$ (see, e.g., [102, Corollary 9.3.1]).

The uniqueness of the optimal solution is a useful property in LO as it allows us to apply the classical sensitivity analysis results. A stronger property plays a similar role in LSIO: an element $\overline{x} \in F$ is a *strongly unique solution* of P if there exists $\alpha > 0$ such that

$$c'x \geq c'\overline{x} + \alpha \|x - \overline{x}\|_2 \text{ for all } x \in F. \tag{1.8}$$

Obviously, (1.8) implies $S = \{\overline{x}\}$.

Theorem 1.1.4 (Uniqueness). $\overline{x} \in F$ *is a strongly unique solution of P if and only if $c \in \text{int } D\left(F; \overline{x}\right)^\circ$.*

So, given $\overline{x} \in F$, if $c \in \text{int } A(\overline{x})$ then $\overline{x}$ is strongly unique, by (1.4), and the converse statement holds if the LFMCQ holds at $\overline{x}$.

The following simple example is used for illustrative purposes throughout this book. Due to the simplicity of its constraints, one can get an explicit (but rather involved) expression of the corresponding marginal function g.

Example 1.1.1. Consider the continuous LSIO problem

$$P : \inf_{x \in \mathbb{R}^2} c'x$$
$$\text{s.t.} \quad -(\cos t) x_1 - (\sin t) x_2 \geq -1, \ t \in \left[0, \frac{\pi}{2}\right],$$
$$x_1 \geq 0 \ (t = 2), \ x_2 \geq 0 \ (t = 3),$$

with different cost vectors c. We have $F = \left\{x \in \mathbb{R}_+^2 : \|x\|_2 \leq 1\right\}$,

$$g(x) = \begin{cases} \max\{\|x\|_2 - 1, -x_1, -x_2\}, & \text{if } x \in \mathbb{R}_+^2, \\ \max\{x_1 - 1, -x_2\}, & \text{if } x \notin \mathbb{R}_+^2 \text{ and } x_2 \leq x_1, \\ \max\{x_2 - 1, -x_1\}, & \text{if } x \notin \mathbb{R}_+^2 \text{ and } x_2 > x_1, \end{cases}$$

and

$$N = K = \text{cone} \left\{ -\begin{pmatrix} \cos t \\ \sin t \\ 1 \end{pmatrix}, t \in \left[0, \frac{\pi}{2}\right]; \begin{pmatrix} 1 \\ 0 \\ 0 \end{pmatrix}, \begin{pmatrix} 0 \\ 1 \\ 0 \end{pmatrix} \right\}$$

(see Fig. 1.1), whose projection on $\mathbb{R}^2$ is $M = \mathbb{R}^2$. Due to the closedness of K, P satisfies the FMCQ and the LFMCQ at any feasible point. Moreover, $\hat{x} = \left(\frac{1}{2}, \frac{1}{2}\right)$ is a Slater point, so that the SCQ and the SSCQ hold too.

(a) If $c = (1, 1)$, $S = \{x^1\}$ with $x^1 = 0_2$ strongly unique. In fact, $D(F; x^1) = A(x^1) = \mathbb{R}_+^2$, with $c \in \text{int } A(\overline{x}) = \mathbb{R}_{++}^2$.

(b) If $c = (-1, -1)$, $S = \{x^2\}$, with $x^2 = \left(\frac{1}{\sqrt{2}}, \frac{1}{\sqrt{2}}\right)$ not strongly unique. Here $D(F; x^2) = \{x \in \mathbb{R}^2 : x_1 + x_2 < 0\} \cup \{0_2\}$ and $A(x^2) = \mathbb{R}_+ c$ (Fig. 1.2 represents both cones translated to x^2).

(c) If $c = (1, 0)$, $S = \{0\} \times [0, 1]$. Let $x^3 = (0, 1) \in S$. Here $D(F; x^3) = \{x \in \mathbb{R}^2 : x_1 \geq 0, x_2 < 0\} \cup \{0_2\}$ and $A(x^3) = \text{cone}\{(0, -1), (1, 0)\}$.

Figure 1.3 represents the graph of the slack functions at x^i, $s(x^i, \cdot)$, $i = 1, 2, 3$.

The existence of strongly unique solution admits a characterization in terms of the relationship between c and K (see Corollary 4.1.1).

Remark 1.1.1 (Antecedents and Extensions). Proofs of Theorems 1.1.1 and 1.1.2 can be found in [102, Theorem 4.4] and [102, Theorem 3.1], respectively. As shown in [71], both results are valid in convex infinite optimization (CIO, in brief). Theorems 1.1.3 and 1.1.4 were first proved in [199] (see also [102, Theorems 7.1 and 10.5]). There exist many versions of the optimality theorem in CIO (see, e.g., [71, 72, 175]).

Fig. 1.1 Characteristic cone

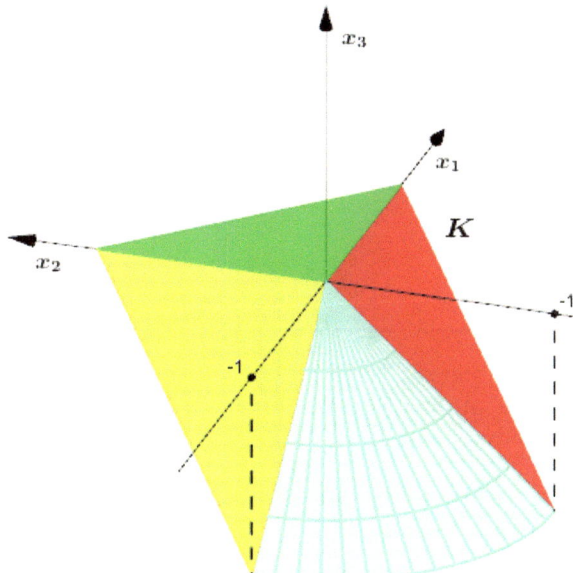

Fig. 1.2 Active and feasible
directions cones

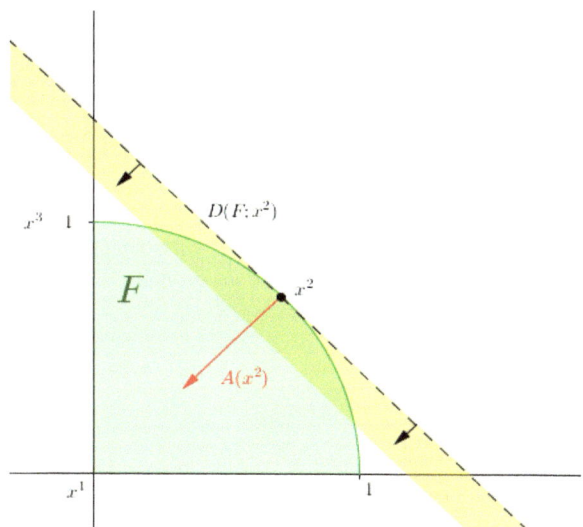

1.2   Duality

If $c \in M$ there exists $\lambda \in \mathbb{R}_+^{(T)}$ such that $\sum_{t \in T} \lambda_t a_t = c$. Then, for any $x \in F$, one has

$$c'x = \sum_{t \in T} \lambda_t a_t' x \geq \sum_{t \in T} \lambda_t b_t. \tag{1.9}$$

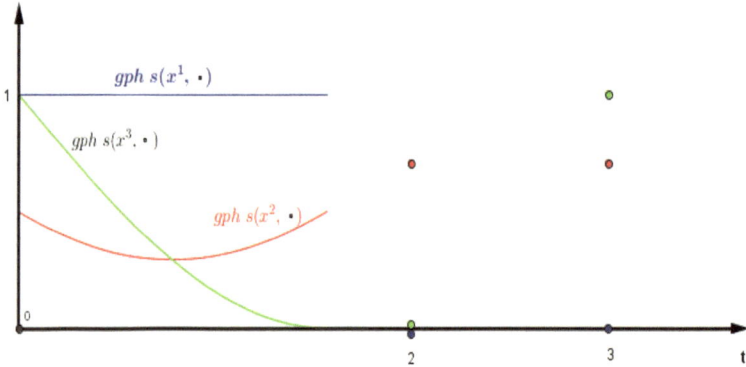

Fig. 1.3 Slack functions of x^i, $i = 1, 2, 3$

The *Haar dual problem* of P consists of maximizing the lower bound for $c'x$ provided by (1.9):

$$D : \sup_{\lambda \in \mathbb{R}_+^{(T)}} \sum_{t \in T} \lambda_t b_t$$
$$\text{s.t.} \quad \sum_{t \in T} \lambda_t a_t = c.$$

We denote by F^D, S^D, and $v(D)$ the *feasible* and *optimal sets* of D, and its *optimal value*, respectively (with the convention that $\sup \emptyset = -\infty$). Notice that D is a LSIO problem as it has finitely many constraints and infinitely many decision variables. By construction, the *weak dual inequality* $v(D) \leq v(P)$ always holds. Observe that $F^D \neq \emptyset$ if and only if $c \in M$.

Other dual LSIO problems can be associated with P following general schemes. For instance, the *Lagrangian dual problem* of P is the unconstrained optimization problem

$$D_L : \quad \sup_{\lambda \in \mathbb{R}_+^{(T)}} \inf_{x \in \mathbb{R}^n} L(x, \lambda).$$

Since

$$\inf_{x \in \mathbb{R}^n} L(x, \lambda) = \inf_{x \in \mathbb{R}^n} \left(\sum_{t \in T} \lambda_t b_t + \left\langle c - \sum_{t \in T} \lambda_t a_t, x \right\rangle \right)$$
$$= \begin{cases} \sum_{t \in T} \lambda_t b_t, & \text{if } \lambda \in F^D, \\ -\infty, & \text{otherwise,} \end{cases}$$

the optimal values and optimal sets of D and D_L coincide, i.e., $v(D) = v(D_L)$ and $S^D = S^{D_L}$.

The pair $P - D$ admits a geometric reformulation in terms of the characteristic cone K. In fact, on one hand, D consists of maximizing the last coordinate, x_{n+1}, on the set

$$\left\{ \sum_{t \in T} \lambda_t (a_t, b_t), \lambda \in \mathbb{R}_+^{(T)} \right\} = \text{cone} \{(a_t, b_t), t \in T\} = N,$$

or, equivalently, on the set $K = N + (0_n, -1)$, under the constraint that $\sum_{t \in T} \lambda_t a_t = c$. So,

$$D_G : \sup_{y \in \mathbb{R}} \{y : (c, y) \in K\}$$

satisfies $v(D) = v(D_G)$ while $S^{D_G} = \{v(D)\}$ whenever $S^D \neq \emptyset$, i.e., $K \cap \{(x, x_{n+1}) \in \mathbb{R}^{n+1} : x = c\}$ is a closed half-line. On the other hand, given $\alpha \in \mathbb{R}$, $\alpha \leq v(P)$ if and only if $c'x \geq \alpha$ is consequence of σ if and only if (by the Farkas Lemma) $(c, \alpha) \in \operatorname{cl} K$. Thus,

$$P_G : \sup_{y \in \mathbb{R}} \{y : (c, y) \in \operatorname{cl} K\}$$

satisfies $v(P) = v(P_G)$ and $S^{P_G} = \{v(P)\}$ whenever $v(P) \in \mathbb{R}$, even though $S = \emptyset$. Of course, the space of decisions is the real line $\mathbb{R}$ for both P_G and D_G. The weak duality $v(D_G) \leq v(P_G)$ is here consequence of the inclusion $K \subset \operatorname{cl} K$ and the optimal sets have at most one element.

The *continuous dual problem* of a continuous LSIO problem P is

$$D_C : \sup_{\mu \in C'_+ (T)} \int_T b_t \, d\mu(t)$$
$$\text{s.t.} \qquad \int_T a_t \, d\mu(t) = c,$$

where $C'_+(T)$ represents the cone of nonnegative regular Borel measures on T. Since $\mathbb{R}_+^{(T)}$ can be seen as the subset of $C'_+(T)$ formed by the nonnegative atomic measures and $P - D_C$ satisfies the weak duality, $v(D) \leq v(D_C) \leq v(P)$. Thus, any condition guaranteeing a zero duality gap for the pair $P - D$ guarantees also a zero duality gap for the pair $P - D_C$ (with attainment of the dual optimal value of D_C whenever $v(D)$ is attained). Thus, we shall consider the Haar dual problem D of P throughout this book.

The main LSIO duality theorems give conditions guaranteeing a zero duality gap with attainment of either the dual or the primal optimal value when $F \neq \emptyset \neq F^D$. These situations are called *strong* (or *infmax*) *duality* and *converse strong* (or *minsup*) *duality*, respectively. The infmax (or strong) duality theorem is a straightforward consequence of the relationship between the pairs $P - D$ and $P_G - D_G$.

Theorem 1.2.1 (Duality). *Let* $F \neq \emptyset \neq F^D$. *Then the following statements hold:*

(i) *If K is closed, then* $v(P) = v(D) \in \mathbb{R}$ *and* $S^D \neq \emptyset$.
(ii) *If $c \in \operatorname{rint} M$, then* $v(P) = v(D) \in \mathbb{R}$ *and S is the sum of a nonempty compact convex set with a linear subspace.*

A feasible pair $(x, \lambda) \in F \times F^D$ is said to be a *complementary solution* of the primal–dual pair $P - D$ if

$$\operatorname{supp} x \cap \operatorname{supp} \lambda = \emptyset,$$

where

$$\operatorname{supp} x := \{t \in T : a'_t x > b_t\} \text{ and } \operatorname{supp} \lambda := \{t \in T : \lambda_t > 0\}$$

are called the *support sets* of x and λ, respectively.

It has been shown in [110] that a pair $(x, \lambda) \in F \times F^D$ is a complementary solution of $P - D$ if and only if $v(D) = v(P)$ and (x, λ) is a primal–dual optimal solution, i.e., $(x, \lambda) \in S \times S^D$. Moreover, given a point $x \in F$, there exists $\lambda \in F^D$ such that (x, λ) is a complementary solution of $P - D$ if and only if x is an optimal solution for some finite subproblem of P. A triplet $(B, N, Z) \in (2^T)^3$ is called an *optimal partition* for P if there exists a complementary solution (x, λ) such that $B = \operatorname{supp} x$, $N = \operatorname{supp} \lambda$, and $Z = T \setminus (B \cup N)$. Then, the nonempty elements of (B, N, Z) form a partition of T (a tripartition when the three sets are nonempty). A partition $(\overline{B}, \overline{N}, \overline{Z})$ is *maximal* if

$$\overline{B} = \bigcup_{x \in S} \operatorname{supp} x, \qquad \overline{N} = \bigcup_{\lambda \in S^D} \operatorname{supp} \lambda \quad \text{and} \quad \overline{Z} = T \setminus (\overline{B} \cup \overline{N}).$$

The uniqueness of the maximal partition is a straightforward consequence of the definition. If there exists an optimal solution pair $(\overline{x}, \overline{\lambda}) \in S \times S^D$ such that $\operatorname{supp} \overline{x} = \overline{B}$ and $\operatorname{supp} \overline{\lambda} = \overline{N}$, then the maximal partition is called the *maximal optimal partition*. Hence, if $S = \{\overline{x}\}$, $S^D = \{\overline{\lambda}\}$, and $v(D) = v(P)$, $\left(\operatorname{supp} \overline{x}, \operatorname{supp} \overline{\lambda}, T \setminus \left(\operatorname{supp} \overline{x} \cup \operatorname{supp} \overline{\lambda}\right)\right)$ is the maximal optimal partition.

If $(\overline{B}, \overline{N}, \overline{Z})$ is an optimal partition such that $\overline{Z} = \emptyset$, then it is a maximal optimal partition. The maximal optimal partition may not exist.

Example 1.2.1. Consider the LSIO problem of Example 1.1.1:

$$P(c) : \inf_{x \in \mathbb{R}^2} c'x$$
$$\text{s.t.} \quad -(\cos t) x_1 - (\sin t) x_2 \geq -1, \ t \in \left[0, \tfrac{\pi}{2}\right],$$
$$x_1 \geq 0 \ (t = 2), \ x_2 \geq 0 \ (t = 3),$$

for three different choices of $c \in \mathbb{R}^2$. We have seen in Example 1.1.1 that K is closed and $M = \mathbb{R}^2$. So, by the duality theorem, $v(P) = v(D)$, with $S \neq \emptyset$ compact and $S^D \neq \emptyset$ for any $c \in \mathbb{R}^2$.

(a) $c = (1, 1)$. We have $S = \{0_2\}$ and, solving the system

$$\left\{\sum_{t \in T} \lambda_t (a_t, b_t) = (c, v(D)), \lambda \in \mathbb{R}_+^{(T)}\right\},$$

we get $S^D = \{\lambda^1\}$, with $\lambda_2^1 = \lambda_3^1 = 1$ and $\lambda_t^1 = 0$ for all $t \in \left[0, \tfrac{\pi}{2}\right]$. Since $\operatorname{supp} 0_2 = \left[0, \tfrac{\pi}{2}\right]$ and $\operatorname{supp} \lambda^1 = \{2, 3\}$, $\left(\left[0, \tfrac{\pi}{2}\right], \{2, 3\}, \emptyset\right)$ is the maximal optimal partition.

(b) $c = (-1, -1)$. Here $S = \{x^2\}$, with $x^2 = \left(\frac{1}{\sqrt{2}}, \frac{1}{\sqrt{2}}\right)$, so that $v(D) = v(P) = -\sqrt{2}$. The uniqueness of D follows from the relationship between S^D and S^{D_G}, and the identification of the characteristic cone K. In fact,

$$K \cap (\{c\} \times \mathbb{R}) = \left\{ \left(-1, -1, -\sqrt{2} - \gamma\right) : \gamma \geq 0 \right\},$$

and $\mathbb{R}_+ \left(-1, -1, -\sqrt{2}\right)$ is an extreme ray of K whose unique generator is $\left(a_{\frac{\pi}{4}}, b_{\frac{\pi}{4}}\right)$. Thus, $S^D = \{\lambda^2\}$, with $\lambda^2_{\frac{\pi}{4}} = \sqrt{2}$ and $\lambda^2_t = 0$ for any $t \neq \frac{\pi}{4}$ (observe that duplicating the constraint corresponding to $t = \frac{\pi}{4}$, nothing changes in the primal problem but S^D would be no longer a singleton). Since $\operatorname{supp} x^2 = T \setminus \{\frac{\pi}{4}\}$ and $\operatorname{supp} \lambda^2 = \{\frac{\pi}{4}\}$, $\left(T \setminus \{\frac{\pi}{4}\}, \{\frac{\pi}{4}\}, \emptyset\right)$ is the maximal optimal partition.

(c) $c = (1, 0)$. Now $S = \{0\} \times [0, 1]$ while $S^D = \{\lambda^3\}$, with $\lambda^3_2 = 1$ and $\lambda^3_t = 0$ otherwise. Here $\operatorname{supp} \lambda^3 = \{2\}$ while, given $x \in S$, we may have $\operatorname{supp} x = [0, \frac{\pi}{2}]$, $\operatorname{supp} x = [0, \frac{\pi}{2}] \cup \{3\}$ or $\operatorname{supp} x = [0, \frac{\pi}{2}[\cup \{3\}$. Thus, the optimal partitions are $\left([0, \frac{\pi}{2}], \{2\}, \{3\}\right)$, $\left([0, \frac{\pi}{2}] \cup \{3\}, \{2\}, \emptyset\right)$, and $\left([0, \frac{\pi}{2}[\cup \{3\}, \{2\}, \{\frac{\pi}{2}\}\right)$ and $\left([0, \frac{\pi}{2}] \cup \{3\}, \{2\}, \emptyset\right)$ turns out to be the maximal optimal partition.

Remark 1.2.1 (Antecedents and Extensions). The seminal papers on Haar's duality were published in the 1960s [59, 60]. As shown in [72], D and D_L are also equivalent to the dual problem in Rockafellar's sense [205], whose feasible set is $\mathbb{R}^{(T)}$.

Proofs of Theorem 1.2.1(i) and its extension to Lagrangian duality in CIO can be found in [102, Theorem 8.2] and [26], respectively. The interest by minsup duality is quite recent. As observed in [107, Remark 6], Theorem 1.2.1(ii) can be proved from [102, Theorem 8.1]. An extension of Theorem 1.2.1(ii) to Lagrangian duality in CIO has been proposed in [107].

1.3 Numerical Methods

We finish this introductory chapter with a brief description of some numerical methods to solve the LSIO problem P in (1.1). The reader cannot expect efficient methods allowing to solve any LSIO problem. Otherwise, we could compute efficiently the optimal value of any optimization problem

$$P_1 : \inf_{x \in X} f(x)$$

by solving the equivalent LSIO problem

$$P_2 : \inf_{y \in \mathbb{R}} -y$$
$$\text{s.t.} \quad y \leq f(x), \ x \in X,$$

with $v(P_1) = -v(P_2)$. More precisely, if $\overline{x}$ is an optimal solution of P_1, then $f(\overline{x})$ is an optimal solution of P_2 and, conversely, if $\overline{y}$ is an optimal solution of P_2, then the optimal set of P_1 is the set of active indices of P_2 at $y = \overline{y}$.

The main drawback with the (linear and nonlinear) semi-infinite methods is the fact that checking the feasibility of a given $\overline{x} \in \mathbb{R}^n$ requires to compute the optimal value, $v(Q(\overline{x}))$, of the so-called *sublevel problem* at $\overline{x} \in \mathbb{R}^n$,

$$Q(\overline{x}) : \inf_{t \in T} s(\overline{x}, t) = \inf_{t \in T} \{ \langle a(t), \overline{x} \rangle - b(t) \},$$

which is a global optimization problem. Even more, some algorithms require the computation of the set $T(\overline{x})$ of global minima of $Q(\overline{x})$, and this is only possible under strong assumptions on s and T, e.g., that $s(\overline{x}, t)$ be a polynomial function of t and T is a finite dimensional interval.

1.3.1   Grid Discretization Methods

Discretization methods generate sequences of points in $\mathbb{R}^n$ converging to an optimal solution of P by solving a sequence of LO problems. These problems are usually subproblems of P of the form

$$P(T_k) : \min_{x \in \mathbb{R}^n} c'x \text{ s.t. } a_t'x \geq b_t \text{ for all } t \in T_k,$$

where T_k is a nonempty finite subset of T for $k = 1, 2, \ldots$ Take a fixed small scalar $\varepsilon > 0$ (called *accuracy*) in order to guarantee finite termination.

Step k : Let T_k be given.

1. Compute a solution x_k of $P(T_k)$.
2. Stop if x_k is feasible within the fixed accuracy ε, i.e., $a_t'x_k \geq b_t - \varepsilon$ for all $t \in T$. Otherwise, replace T_k with a new grid T_{k+1}.

Obviously, x_k is unfeasible before optimality. Grid discretization methods select a priori sequences of *grids* $(T_k)_{k=0}^{\infty}$ (usually satisfying $T_k \subset T_{k+1}$ for all k). The alternative discretization approaches generate the sequence $(T_k)_{k=0}^{\infty}$ inductively. For instance, the classical Kelley cutting plane approach consists of taking $T_{k+1} = T_k \cup \{t_k\}$, for some $t_k \in T$ (as in Fig. 1.4), or $T_{k+1} = (T_k \cup \{t_k\}) \setminus \{t_k'\}$ for some $t_k' \in T_k$ (if an elimination rule is included).

Convergence of discretization methods requires P to be continuous. This guarantees the convergence in the case of the Kelley cutting plane method while

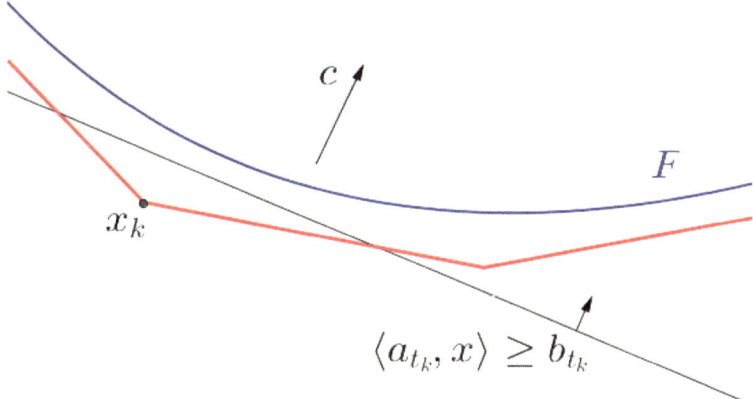

Fig. 1.4 Feasibility cut

grid discretization methods require in addition the following *density assumption* on P (actually on T): T is the union of a finite set U with another compact set V such that $V = \mathrm{cl}(\mathrm{int}\, V)$. In that case, the sequence of grids should satisfy $U \subset T_k$ for all k.

Concerning the latter requirement, consider the problems in Example 1.1.1, which are continuous and $T = \left[0, \frac{\pi}{2}\right] \cup \{2, 3\}$ satisfies the density assumption with $U = \{2, 3\}$ and $V = \left[0, \frac{\pi}{2}\right]$. If T_k is a grid in T such that $U \subsetneq T_k$, then the feasible set F recedes in at least one of the directions $(-1, 0)$ or $(0, -1)$ and $P(T_k)$ is unbounded in case (a), so that the sequence $(x_k)_{k=0}^{\infty}$ may not exist. In the contrary, requiring $U \subset T_k$ for all k, we have $x_k = 0_2$ for k and $0_2 = \lim_{k \to \infty} x_k \in S$.

The main drawbacks with these methods are undesirable jamming in the proximity of S (unless P has a strongly unique optimal solution) and the increasing size of the auxiliary problems $P(T_k)$ (unless elimination rules are implemented). These methods are only efficient for low-dimensional index sets, i.e., T is contained in some Euclidean space and $\dim \mathrm{aff}\, T \le 3$ (otherwise the *cardinality* $|T_k|$ of T_k grows very fast with k). For instance, if $T = \prod_{i=1}^{m} [\alpha_i, \beta_i]$ and T_k is formed by successive bipartitions of the intervals $[\alpha_i, \beta_i]$, $i = 1, \ldots, m$, then $|T_{k+1}| = \left(2^k + 1\right)^m > 2^{km}$ for $k = 1, 2, \ldots$ For more details see [102, Chap. 11] and [178], and references therein.

1.3.2 Central Cutting Plane Methods

Central cutting plane methods start each step with a polytope containing a sublevel set of P and calculate a certain *center* of this polytope. The polytope is then updated by aggregating to its defining system either a feasibility cut (if the center is unfeasible) or an objective cut (otherwise). Let $\varepsilon > 0$ be a fixed accuracy.

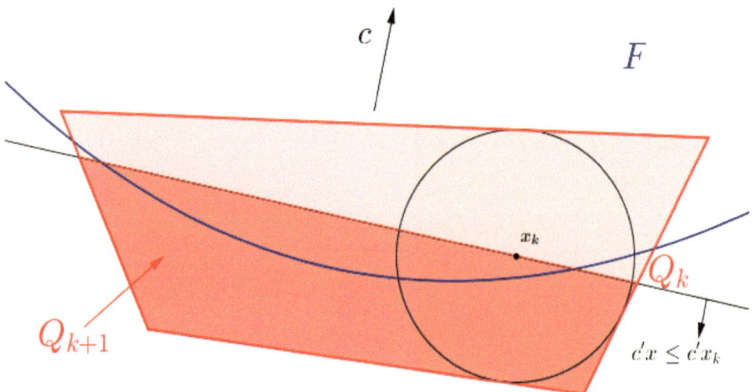

Fig. 1.5 Objective cut

Step k : Let Q_k be a polytope containing some sublevel set of P.

1. Compute a center x_k of Q_k.
2. If $x_k \notin F$, set $Q_{k+1} = \{x \in Q_k : a'_t x \geq b_t\}$, where $t \in T$ satisfies $a'_t x_k < b_t$, and $k = k + 1$. Otherwise, continue.
3. If $c' x_k \leq \min_{x \in Q_k} c' x + \varepsilon$, stop. Otherwise, set $Q_{k+1} = \{x \in Q_k : c' x \leq c' x_k\}$ and $k = k + 1$.

Concerning item 1, an obvious condition for the existence of Q_0 is the boundedness of S. Now we assume that $Q_k = \{x \in \mathbb{R}^n : c'_i x \geq d_i, i \in I\}$, with I finite and $c_i \neq 0_n$ for all $i \in I$. If one chooses the *geometric center* (as in [21]), the radius of the greatest ball for the norm $\|\cdot\|_2$ contained in Q_k is $\max_{x \in Q_k} \min_{i \in I} d_2(x, H_i)$, where $H_i := \{x \in \mathbb{R}^n : c'_i x = d_i\}$. Since $d_2(x, H_i) = \frac{c'_i x - d_i}{\|c_i\|_2}$, $i \in I$, this auxiliary problem is equivalent to $\min_{x \in Q_k} \max_{i \in I} -d_2(x, H_i)$ and also to the LO problem

$$\min_{(x, y) \in \mathbb{R}^{n+1}} y$$
$$\text{s.t.} \qquad c'_i x - d_i + \|c_i\|_2 \, y \geq 0, i \in I,$$
$$c'_i x \geq d_i, i \in I.$$

Alternatively, if one chooses the *analytic center* (as in [192]), the auxiliary problem consists of computing a global minimizer of $\inf_{x \in \mathbb{R}^n} f(x)$, where f is a *barrier function* for Q_k (i.e., a function f such that $f(x) \to +\infty$ as $x \to \text{bd } Q_k$), e.g., the *logarithmic barrier function* $f(x) = -\sum_{i \in I} \log(c'_i x - d_i)$, if $x \in \text{int } Q_k$, and $f(x) = +\infty$, otherwise.

Item 2 aggregates to the current polytope Q_k a feasibility cut (some constraint of P violated by x_k) when x_k is unfeasible, whereas item 3 checks the ε-optimality of x_k, aggregating to Q_k an objective cut when the result is negative. Figure 1.5 illustrates an objective cut.

Typically, these methods generate sequences of feasible and unfeasible points which provide stopping rules for ε-optimality and stop before optimality at a feasible solution. For instance, the method proposed in [65] generates at each step a non-feasible point together with a feasible one, the result of shifting the current unfeasible point toward a fixed Slater point with step length easily computable.

In particular, the so-called interior point constraint generation algorithm in [192], inspired in [182], updates the current discretization P_k (interpreted as a dual problem of certain LO problem in standard format, say D_k) of P by selecting a point in the vicinity of the central path of P_k and aggregating to the constraints of P_k some violated constraints; then the full dimension of the new feasible set F_{k+1} is recovered and the central path is updated. This process continues until the barrier parameter is small enough, i.e., the duality gap approaches to zero. This algorithm does not generate feasible points of P (so that it is not an interior point method) but converges to an ε-optimal solution after a finite number of constraints is generated. Assuming that P is continuous and F is full dimensional and bounded, the authors provide complexity bounds on the number of Newton steps needed and on the total number of constraints that is required for the overall algorithm to converge.

Discretization and cutting plane methods share the above-mentioned drawbacks. Convergence of central cutting plane methods requires continuity of P together with the boundedness of S, i.e., $c \in \text{int } M$. Additionally, the interior point constraint generation algorithm requires the stronger assumption that F is a convex body (i.e., a compact convex full dimensional set), i.e., the pointedness of cl K (i.e., that cl K has no lines). Recall that the cones M and cl K are defined in terms of the data. All these conditions hold in the problems considered in Example 1.1.1.

In conclusion, efficient implementations of the cutting plane methods turn out to be computationally faster than the discretization by grids counterparts, but they require stronger assumptions.

1.3.3 Reduction Methods

Reduction methods, which were already known from Chebyshev approximation, replace P with a nonlinear system obtained from the optimality conditions.

Under suitable conditions (recall Theorem 1.1.3), if $\overline{x}$ is a minimizer of P, there exist indices $\overline{t}_j \in T(\overline{x})$, $j = 1, \ldots, q(\overline{x})$, with $q(\overline{x}) \in \mathbb{N}$ depending on $\overline{x}$, and nonnegative multipliers $\overline{\lambda}_j$, $j = 1, \ldots, q(\overline{x})$, such that

$$c = \sum_{j=1}^{q(\overline{x})} \overline{\lambda}_j a\left(\overline{t}_j\right).$$

We assume also the availability of a description of $T \subset \mathbb{R}^m$ as

$$T = \{t \in \mathbb{R}^m : u_i(t) \geq 0, \ i = 1, \ldots, m\}, \tag{1.10}$$

where u_i is smooth for all $i = 1, \ldots, m$.

Observe that $q(\bar{x})$ is the number of global minima of the sublevel problem $Q(\bar{x})$, provided that $v(Q(\bar{x})) = 0$. In that case, $\bar{t}_j \in T(\bar{x})$ if and only if $\bar{t}_j$ is a minimizer of the (finite) *sublevel problem* at $\bar{x}$

$$Q(\bar{x}) : \inf_{t \in T} \langle a(t), \bar{x} \rangle - b(t)$$
$$\text{s.t.} \quad u_i(t) \geq 0, i = 1, \ldots, m.$$

Then, under some constraint qualification, the classical KKT theorem yields the existence of nonnegative multipliers $\bar{\theta}_i^j$, $i = 1, \ldots, m$, such that

$$\langle \nabla_t a(\bar{t}_j), \bar{x} \rangle - \nabla_t b(\bar{t}_j) = \sum_{i=1}^m \bar{\theta}_i^j \nabla_t u_i(\bar{t}_j) \qquad (1.11)$$

and

$$\bar{\theta}_i^j u_i(\bar{t}_j) = 0, \; i = 1, \ldots, m.$$

In the typical case that T is an interval $[\alpha, \beta] \subset \mathbb{R}$, $m = 2$, $u_1(t) = t - \alpha$, $u_2(t) = \beta - t$, and (1.11) reads

$$\left\langle \frac{da(\bar{t}_j)}{dt}, \bar{x} \right\rangle - \frac{db(\bar{t}_j)}{dt} = \sum_{i=1}^m \bar{\theta}_i^j \nabla_t u_i(\bar{t}_j).$$

Step k : Start with a given x_k (not necessarily feasible).

1. Estimate $q(x_k)$.
2. Apply N_k steps of a quasi-Newton method (for finite systems of equations) to

$$\begin{cases} c = \displaystyle\sum_{j=1}^{q(x_k)} \lambda_j a(t_j) \\ \langle a(t_j), x \rangle = b(t_j), \; j = 1, \ldots, q(x_k) \\ \langle \nabla_t a(t_j), x \rangle - \nabla_t b(t_j) = \displaystyle\sum_{i=1}^m \theta_i^j \nabla_t u_i(t_j), \; j = 1, \ldots, q(x_k) \\ \theta_i^j u_i(t_j) = 0, i = 1, \ldots, m, \; j = 1, \ldots, q(x_k) \end{cases} \qquad (1.12)$$

(with unknowns x, t_j, λ_j, θ_i^j, $i = 1, \ldots, m$, $j = 1, \ldots, q(x_k)$), leading to iterates $x_{k,l}, l = 1, \ldots, N_k$.
3. Set $x_{k+1} = x_{k,N_k}$ and $k = k + 1$.

The reduction methods have two drawbacks: first, they require strong assumptions on P (smoothness of the functions describing F and T), the compactness of T, the existence of some Slater point (so that P satisfies FMCQ); and, second, they require sufficiently accurate approximate solution of (1.12) to start. The advantage of reduction methods is their fast asymptotic behavior (as the quasi-Newton method they use). A reasonable strategy consists of combining, in a two-phase method, the advantages of discretization, which may provide an estimation of $q(x_k)$ and an approximate solution of (1.12), and reduction, to improve this approximate solution. Unfortunately, no theoretical result supports the decision to go from phase 1 (discretization) to phase 2 (reduction) and, in practice, it is difficult to compute a suitable starting point for phase 2.

Example 1.3.1. Let us apply the above methods to the simple LSIO problem of Example 1.1.1(b):

$$P : \inf_{x \in \mathbb{R}^2} -x_1 - x_2$$
$$\text{s.t.} \quad -(\cos t) x_1 - (\sin t) x_2 \geq -1, \ t \in \left[0, \frac{\pi}{2}\right],$$
$$x_1 \geq 0 \ (t = 2), \ x_2 \geq 0 \ (t = 3).$$

Take the sequence of grids $T_k = \{2, 3\} \cup \{\frac{\pi i}{4k} : i = 0, \dots, 2k\}, k \in \mathbb{N}$. Then $\frac{\pi}{4} \in T_k$ for all $k \in \mathbb{N}$ and the unique optimal solution of $P(T_k)$ is $x_k = \left(\frac{1}{\sqrt{2}}, \frac{1}{\sqrt{2}}\right)$, with $a_t' x_k \geq b_t - \varepsilon$ for all $t \in T$. Thus, the discretization algorithm, with an arbitrary tolerance $\varepsilon > 0$, stops at step 1 with $x_1 \in S$. If, alternatively, we take $T_k = \{2, 3\} \cup \{\frac{\pi i}{4k+2} : i = 0, \dots, 2k\}, k \in \mathbb{N}$, the sequence of optimal solutions generated by the grid discretization method is

$$x_k = \left[\sin\left(\left(\frac{4k+1}{8k+4}\right)\pi\right)\right]^{-1} \left(\frac{1}{\sqrt{2}}, \frac{1}{\sqrt{2}}\right) \rightarrow \left(\frac{1}{\sqrt{2}}, \frac{1}{\sqrt{2}}\right) \in S,$$

with $\|x_k\|_2 > 1$ for all k, so that the algorithm terminates after a finite number of steps with some unfeasible point.

Now we apply the geometric central cutting plane method. Taking $Q_0 = [0, 1]^2 \supset F$, $x_0 = \left(\frac{1}{2}, \frac{1}{2}\right) \in F$. So the 1st cut is an objective one and $Q_1 = \{x \in \mathbb{R}^2 : x_1 + x_2 \geq 1, x_1 \geq 0, x_2 \geq 0\}$, whose geometric center is its incenter

$$x_1 = \left(\frac{\sqrt{2}+1}{\sqrt{2}+2}, \frac{\sqrt{2}+1}{\sqrt{2}+2}\right) = \left(\frac{1}{\sqrt{2}}, \frac{1}{\sqrt{2}}\right) \in S,$$

and the algorithm terminates in one step independently of the chosen tolerance.

If one applies the analytic central cutting plane method with logarithmic barrier function and the same initial polytope as above, $Q_0 = [0, 1]^2$, x_0 and Q_1 are the same as before, but now $x_1 = \left(\frac{2}{3}, \frac{2}{3}\right) \in F$. Then, $Q_2 = \{x \in \mathbb{R}^2 : x_1 + x_2 \geq \frac{4}{3}, x_1 \geq 0, x_2 \geq 0\}$, $x_2 = \left(\frac{7}{9}, \frac{7}{9}\right) \notin F$ and so the next cut is a feasibility cut. Since the deepest cut corresponds to $t = \frac{\pi}{4}$, $Q_3 = \{x \in \mathbb{R}^2 : \sqrt{2} \geq x_1 + x_2 \geq \frac{4}{3}, x_1 \geq 0, x_2 \geq 0\}$, $x_3 = (0.68559, 0.68559) \in F$, and so on. Observe that this algorithm generates feasible and unfeasible iterates so that termination can be produced at a feasible approximate solution.

Concerning the reduction approach, we must represent $T = \left[0, \frac{\pi}{2}\right] \cup \{2, 3\}$ as in (1.10), e.g., $T = \{t \in \mathbb{R} : u(t) \geq 0\}$ with $u(t) = -t \left(t - \frac{\pi}{2}\right)(t - 2)^2 (t - 3)^2$, and estimating the number of active indices at the minimum, in our case 1 or 2 (the maximum number of active indices at the boundary of F). Taking $q(x) = 1$, (1.12) becomes

$$
\begin{cases}
(1, 1) = \lambda(\cos t, \sin t) \\
(\cos t) x_1 + (\sin t) x_2 = 1 \\
(\sin t) x_1 - (\cos t) x_2 = \theta \frac{du}{dt} \\
\theta u(t) = 0
\end{cases}
\tag{1.13}
$$

It is easy to check that $(x_1, x_2, t, \lambda, \theta) = \left(\frac{1}{\sqrt{2}}, \frac{1}{\sqrt{2}}, \frac{\pi}{4}, 2, 0\right)$ is a solution of (1.13). Newton and quasi-Newton methods provide sequences in $\mathbb{R}^5$ converging fast (at least superlinearly) to this point provided an approximate solution is available (unfortunately, getting it is a difficult task!).

1.3.4 Feasible Point Methods

Feasible point methods generate sequences $(x_k)_{k=1}^{\infty}$ of feasible points such that the corresponding sequence of images by the objective function $(c'x_k)_{k=1}^{\infty}$ are non-increasing. The main drawback with these methods is the computational effort required to find the optimal set of the sublevel problem at the current iterate x_k:

$$
Q(x_k) : \min_{t \in T} s(x_k, t) = \min_{t \in T} \{\langle a(t), x_k\rangle - b(t)\}.
$$

In fact, the computation of all the global minima of $Q(x_k)$ is only possible whenever T is a compact interval in $\mathbb{R}$ and $a_1(\cdot), \ldots, a_n(\cdot), b(\cdot) \in C^{\infty}(T)$ (the class of analytic functions on T, which contains the polynomial functions), in which case P is said to be *analytic*. Classical feasible point methods generate a feasible direction at the current iterate x_k by solving a certain LO problem, the next iterate being the result of performing a linear search in this direction improving the objective as much as possible until getting a point $x_{k+1} \in \text{bd } F$. The simplex-like algorithm of Anderson and Lewis [5] consists of alternating purification steps (providing an extreme point of F from the current iterate) and line search steps providing a point of bd F. This method was adapted in [169] to problems which are analytic by blocks. The convergence of the sequences generated by simplex-like methods to an optimal solution of P is not guaranteed. Stein and Still [215] have proposed an interior point method for (not necessarily linear) semi-infinite programs whose LSIO version requires the constraint system to be formed by subsystems of the form $\{\langle a(t), x\rangle \geq b(t), \ t \in T\}$, where T is a convex subset of some vector space, and the function $t \mapsto \langle a(t), x\rangle - b(t)$ to be affine (a strong assumption).

More recently, Floudas and Stein [83] developed an efficient feasible point method whose underlying idea consists of replacing the hard global auxiliary problem $Q(x_k)$ by a suitable convexification under assumptions weaker than those of [169] (the coefficients of the inequalities of each block must be $\mathcal{C}^2$ instead of $\mathcal{C}^\infty$).

Remark 1.3.1 (Simplex-Like Methods). In the seminal papers [59, 60] it was observed that the extreme points of F^D could be characterized in an algebraic way (as in LO). The corresponding simplex method for D was described in algebraic terms in [90] and in geometric terms in [98]. An extension to arbitrary infinite dimensional spaces has been proposed in [212].

Simplex methods for P are only possible under strong assumptions on the constraints. A purification method for analytic LSIO problems was proposed in [5]. In the same paper was proposed the so-called hybrid method, which alternates purification steps with linear search steps, providing infinite sequences in F with non-increasing images by the objective function. A simplex method was proposed in [6] for a class of LSIO problems whose feasible set is quasipolyhedral (i.e., a set whose intersection with polytopes is either empty or a polytope). The advantage of simplex-like methods are their generality (the dual simplex method does not require continuity) and their common drawbacks are the lack of convergence theorems verifiable in practice and the need of exact solutions of the subproblems (an unrealistic requirement).

In the absence of continuity, the LSIO problems could also be approximately solved by a suitable adaptation of the stochastic approach of [33].

Remark 1.3.2 (Available Solvers). The unique publicly available software for (linear and nonlinear) semi-infinite optimization, NSIPS,[1] uses the SIPAMPL software package, which extends AMPL environment to the SIPAMPL database (see [223] and the SIPAMPL manual[2] for additional information). NSIPS is available on the NEOS server[3] and includes four solvers: a discretization solver, a penalty solver, a sequential quadratic programming solver, and an unfeasible quasi-Newton interior point solver. Another family of publicly available solvers for (linear and convex) semi-infinite optimization is in preparation on the basis of the Remez penalty smoothing algorithm in [8] and its implementation in [7]. Simple descriptions of the most efficient LSIO methods (also for nonlinear semi-infinite optimization problems), together with their corresponding comments, have been uploaded to NEOS Optimization Guide[4] by one of the authors.

Concerning commercial software, the Optimization Toolbox of Matlab version 2 contains a solver for semi-infinite optimization with either $T \subset \mathbb{R}$ or $T \subset \mathbb{R}^2$, called

---

[1] http://www.norg.uminho.pt/aivaz/nsips.html

[2] http://plato.la.asu.edu/ftp/sipampl.pdf

[3] http://www.neos-server.org/neos/

[4] http://www.neos-guide.org/algorithms

fseminf (*seminf* in the version 1.5), whose use is described at the Matlab Tutorial.[5] The LSIO problems are solved by *fseminf* via discretization.

Remark 1.3.3 (LSIO Applications). LSIO has been used: first, as a conceptual tool in economic theory, games, or geometry; second, as a computational tool in functional approximation, robust statistics, or semidefinite optimization; and third, as a modeling tool for real problems arising in engineering, health care, or spectrometry. Many users have had difficulties with the latter type of applications due to the lack, until recently, of publicly available software (usually the authors have felt obliged to implement standard or ad hoc numerical methods). We enumerate below some fields where LSIO has been applied in at least one of the three ways. A large collection of references, published before 2010, can be found in http://wwwhome.math.utwente.nl/~stillgj/sip/lit-sip.pdf.

- Environmental engineering [90, 102, 125, 126, 129, 142, 237].
- Optimal design [155, 231].
- Telecommunication networks [91].
- Control problems [90, 91].
- Economic theory [91].
- Finance [65, 162, 167, 186, 196, 224].
- Game theory [91].
- Spectrometry [61].
- Health care [192].
- Probability and Statistics [2, 22, 23, 80, 81, 91, 102].
- Machine Learning [18, 184, 185, 193, 194, 213, 217].
- Data envelopment analysis [91, 102].
- Functional approximation [90, 102, 132, 163, 218].
- Computational linear algebra [102].
- Linear functional equations [90, 102].
- Convex geometry [91, 149, 198].
- Location problems [102].
- Robust optimization [91, 178].
- Semidefinite optimization [91, 164].
- Geometric optimization [102].
- Combinatorial optimization [164].

Remark 1.3.4 (Two Open Problems in Deterministic LSIO).

1. Convergence theorems for simplex-like methods.
2. Complexity analysis of LSIO methods.

---

[5]http://serdis.dis.ulpgc.es/~ii-its/MatDocen/laboratorio/manuales/OPTIM_TB.PDF

Chapter 2
Modeling Uncertain Linear Semi-infinite Optimization Problems

In most LSIO applications part of the data, if not all of them, are uncertain as a consequence of error measurements or estimations. This uncertainty is inherent to the data in fields as environmental engineering, telecommunications, finance, spectrometry, health care, statistics, machine learning, or data envelopment analysis, just to mention some applications listed in Remark 1.3.3. In this chapter we consider given an uncertain LSIO problem

$$P_0 : \inf_{x \in \mathbb{R}^n} c'x$$
$$\text{s.t.} \quad a_t'x \geq b_t, \ t \in T, \tag{2.1}$$

which is many times the result of perturbing the data of a nominal problem

$$\overline{P} : \inf_{x \in \mathbb{R}^n} \overline{c}'x$$
$$\text{s.t.} \quad \overline{a}_t'x \geq \overline{b}_t, \ t \in T. \tag{2.2}$$

Nevertheless, most authors ignore this fact, limiting themselves to solve a particular instance of P_0 without further analysis. As uncertain optimization problems can only be analyzed and/or solved in the framework of specific models, this chapter is intended to help potential users of LSIO to choose suitable models through the description and comparison of different alternatives and the discussion of a significant uncertain LSIO problem. The next section introduces five different paradigms to treat uncertainty: the stochastic, the fuzzy, the interval, the robust, and the parametric approaches. Almost all the existing literature on uncertain LSIO is focused on the latter two approaches.

M.A. Goberna and M.A. López, *Post-Optimal Analysis in Linear Semi-Infinite Optimization*, SpringerBriefs in Optimization, DOI 10.1007/978-1-4899-8044-1_2, © Miguel A. Goberna, Marco A. López 2014

2.1   Five Paradigms to Treat Uncertain LSIO Problems

2.1.1   The Stochastic Approach

Assume that the uncertain data are random variables with a known probability distribution. Each realization of these random variables provides a deterministic LSIO problem called *scenario*. Although it is impossible to obtain in practice the probability distribution of the optimal value, its empirical distribution can be approximated via simulation by solving a sample of scenarios.

Probabilistic models are a subclass of stochastic models consisting of optimization problems with a deterministic objective function and constraints involving probabilities of events which are expressed in terms of the random data. Fuzzy models and interval models can be seen as variants of the stochastic models.

2.1.2   The Fuzzy Approach

Now we recall some standard concepts on single-valued functions which are used in the fuzzy and the parametric approaches. Let X be a topological space and denote by $\mathfrak{N}_x$ the family of neighborhoods of $x \in X$. Let $f : X \to \overline{\mathbb{R}}$ be given. Recall that a function f which is finite-valued around $\overline{x} \in X$ is continuous at that point if, for each $\varepsilon > 0$, there exists $V \in \mathfrak{N}_x$ such that $f(\overline{x}) - \varepsilon < f(x) < f(\overline{x}) + \varepsilon$ for all $x \in V$. This concept can be split into two weaker ones:

- f is *lower semicontinuous* at $\overline{x} \in X$ (lsc in brief) if, for each $\lambda < f(\overline{x})$ there exists $V \in \mathfrak{N}_x$ such that $\lambda < f(x)$ for all $x \in V$; f is lsc if it is lsc at any $x \in X$.
- f is *upper semicontinuous* at $\overline{x} \in X$ (usc) if, for each $\lambda > f(\overline{x})$ there exists $V \in \mathfrak{N}_x$ such that $\lambda > f(x)$ for all $x \in V$; f is usc if it is usc at any $x \in X$.

It is easy to prove that f is lsc if and only if epi f is closed. This is the case of the marginal function g defined in (1.3) as epi g is intersection of closed half-spaces. The *lsc hull* of f is the greatest lsc minorant of f, i.e., the function whose epigraph is cl epi f. Obviously, f is usc at $\overline{x} \in X$ if and only if its opposite function $-f$ is lsc at $\overline{x}$. So, f is usc if and only if hypo f is closed and the *usc hull* of f (i.e., the smallest usc majorant of f) is the function whose hypograph is cl hypo f. Moreover, if f is finite-valued, it is continuous if and only if epi f and hypo f are closed, and this implies that gph f is closed. As the function $f : \mathbb{R} \to \mathbb{R}$ such that

$$f(x) = \begin{cases} 0, & \text{if } x = 0, \\ \frac{1}{x^2}, & \text{otherwise,} \end{cases} \tag{2.3}$$

shows, the converse is not true unless $f : \mathbb{R}^n \to \mathbb{R}$ is bounded.

Now we assume that $X = \mathbb{R}^n$. Recall that f is *concave* (proper, *quasiconcave*) when $-f$ is convex (proper, quasiconvex, respectively), i.e., when hypo f is convex (hypo f is a nonempty set without vertical lines, the superlevel sets of f are convex, respectively). We define the *domain* of a concave function f as the domain of the convex function $-f$, i.e., the convex set dom $f := \{x \in \mathbb{R}^n : f(x) > -\infty\}$. Analogously, we define the *concave subdifferential* of f at $\overline{x} \in \mathbb{R}^n$ as the symmetric w.r.t. the origin of the convex subdifferential of the convex function $-f$, i.e., the closed convex set

$$\partial f(\overline{x}) := \{u \in \mathbb{R}^n : f(x) \leq f(\overline{x}) + u'(x - \overline{x}) \ \forall x \in \mathbb{R}^n\}.$$

A *fuzzy set* A defined in a topological space X (called *universal set*) is characterized by a usc function $\mu_A : X \to [0, 1]$ called *membership function* of A whose images $\mu_A(x), x \in X$, represent the grade of membership of x in A (μ_A can be seen as the fuzzy counterpart of a probability distribution on X). A *fuzzy number* A is a fuzzy set defined in $\mathbb{R}$ whose membership function is quasiconcave and satisfies $\sup_{x \in \mathbb{R}} \mu_A(x) = 1$.

Denote by $\mathfrak{F}$ the class of fuzzy numbers. The product αA of $\alpha \in \mathbb{R}$ and $A \in \mathfrak{F}$ is the fuzzy number corresponding to the membership function $\mu_{\alpha A}(x) := \mu_A\left(\frac{x}{\alpha}\right)$, if $\alpha \neq 0$, and μ_{0A} is the *characteristic function* χ^0 of 0 (i.e., $\chi^0(0) = 1$ and $\chi^0(x) = 0$ otherwise) while the sum $A + B$ of $A, B \in \mathfrak{F}$ is the fuzzy number whose membership function is

$$\mu_{A+B}(x) := \min_{x = u + v} \{\mu_A(u), \mu_B(v)\}.$$

So, the linear combination $a'x = \sum_{1=1}^{n} a_i x_i$ of $a_1, \ldots, a_n \in \mathfrak{F}$, with $x_1, \ldots, x_n \in \mathbb{R}$, is well defined by induction. Observe that, by definition of fuzzy number, given $A \in \mathfrak{F}$ and a level $\lambda \in [0, 1]$, the superlevel set $\{x \in \mathbb{R} : \mu_A(x) \geq \lambda\}$ is a closed convex set; so, we can write $\{x \in \mathbb{R} : \mu_A(x) \geq \lambda\} = [A_\lambda^l, A_\lambda^u]$, where $A_\lambda^l, A_\lambda^u \in \overline{\mathbb{R}} := \mathbb{R} \cup \{\pm\infty\}$. The fuzzy numbers can be compared in different ways. The classical one consists of defining $A \succeq B$ whenever $A_\lambda^l \geq B_\lambda^l$ and $A_\lambda^u \geq B_\lambda^u$ for all $\lambda \in [0, 1]$, but then $\succeq$ is just a partial order on $\mathfrak{F}$. Total binary relations can be defined on $\mathfrak{F}$ by means of centrality measures of the membership functions called *ranking functions*. For instance, the *Roubence ranking function* $R(A) = \frac{1}{2} \int_0^1 \left(A_\lambda^l + A_\lambda^u\right) d\lambda$ (an average of the centers of the superlevel sets of μ_A) allows us to define $A \succeq B$ whenever $R(A) \geq R(B)$; since $R(A_\alpha) = 0$ for any $A_\alpha \in \mathfrak{F}$ such that $\mu_{A_\alpha}(x) = \exp\left(-\alpha x^2\right)$, $\alpha > 0$, we conclude that the latter binary relation $\succeq$ is reflexive and transitive, but not antisymmetric.

Thus, once selected a suitable binary relation $\succeq$ on $\mathfrak{F}$, the linear inequality $a'x \succeq b$, where $a_1, \ldots, a_n, b \in \mathfrak{F}$, has a precise meaning. Now the problem is how to replace a fuzzy inequality $a'x \succeq b$ with a (possibly nonlinear) deterministic system (crisp system, in fuzzy terminology), taking into account that generally $(A + B)_\lambda^l \neq$

$(A)^l_\lambda + (B)^l_\lambda$, $(A + B)^u_\lambda \neq (A)^u_\lambda + (B)^u_\lambda$, and $R(A + B) \neq R(A) + R(b)$. This can be done by defining an ad hoc membership function for $a'x$. The fuzzy approach sketched here interprets the uncertain scalar data in P_0 as fuzzy numbers with known membership functions, and the task consists of computing a vector $\overline{x} \in F$ (the solution set of the fuzzy system $\{a'_t x \succeq b_t, \ t \in T\}$) such that the fuzzy inequality $c'x \succeq c'\overline{x}$ holds for all $x \in F$. The latter fuzzy system is frequently inconsistent when one compares the fuzzy numbers by means of the partial order defined through the superlevel sets.

2.1.3 The Interval Approach

Assume that any instance of the uncertain scalar data in P_0 takes values on a given interval (this assumption is weaker than the uniform distribution on the corresponding interval). In other words, each scalar data in P_0 is interpreted as an interval $A^\pm = [A^l, A^u]$, $A^l, A^u \in \mathbb{R}$.

Denote by $\mathfrak{I}$ the class of compact intervals in $\mathbb{R}$. We define the product of $\alpha \in \mathbb{R}$ times $A^\pm \in \mathfrak{I}$ by $\alpha A^\pm := [\alpha A^l, \alpha A^u]$, if $\alpha \geq 0$, and $\alpha A^\pm := [\alpha A^u, \alpha A^l]$, otherwise. Similarly, we define the sum of $A^\pm, B^\pm \in \mathfrak{I}$ as $A^\pm + B^\pm := [A^l + B^l, A^u + B^u]$. This way, given $a^\pm_1, \dots, a^\pm_n \in \mathfrak{I}$ and $x_1, \dots, x_n \in \mathbb{R}$, the linear combination $\sum_{i=1}^n x_i a^\pm_i$ is a well-defined element of $\mathfrak{I}$. Defining $A^\pm \succeq B^\pm$ whenever $A^l \geq B^l$ and $A^u \geq B^u$, $\succeq$ defines a partial order on $\mathfrak{I}$, so that the inequalities $\sum_{i=1}^n x_i a^\pm_i \succeq b^\pm$ and $\sum_{i=1}^n x_i c^\pm_i \succeq \sum_{i=1}^n \overline{x}_i c^\pm_i$ have precise meanings too.

The task consists then of determining a vector $\overline{x} \in F$ (the solution set of the interval system $\{\sum_{i=1}^n x_i a^\pm_{it} \succeq b^\pm_t, \ t \in T\}$) such that $\sum_{i=1}^n x_i c^\pm_i \succeq \sum_{i=1}^n \overline{x}_i c^\pm_i$ for all $x \in F$. In this model the optimal value $\sum_{i=1}^n \overline{x}_i c^\pm_i$ is also an interval.

2.1.4 The Robust Approach

This approach provides a deterministic framework for studying mathematical programming problems under uncertainty. Robust optimization models are based on the description of uncertainty via sets, as opposed to probability distributions (membership functions, intervals) which are used in stochastic (fuzzy, interval) approaches.

Robust models assume that all instances of the data belong to prescribed sets (not necessarily intervals or boxes), but now the task consists of minimizing the worst possible value of the objective function on the set of points which are feasible for any possible instance of the constraints. Indeed, when only the cost vector c is uncertain in P_0, and its uncertainty set is $C \subset \mathbb{R}^n$, then the robust model replaces the task "inf $c'x$" in P_0 with "inf $\sup_{c \in C} c'x$"; alternatively, when the constraint corresponding to index $t \in T$ is uncertain, with uncertainty set $U_t \subset \mathbb{R}^{n+1}$, then the robust model replaces $a'_t x \geq b_t$ in P_0 with the linear semi-infinite system $\{a'x \geq b, (a, b) \in U_t\}$; consequently, when the uncertainty affects both costs and constraints, defining $U_t := \{(a_t, b_t)\}$ whenever the constraint $a'_t x \geq b_t$ is certain, risk-averse decision makers prefer the (pure) robust optimization model

$$P_R : \inf_{x \in \mathbb{R}^n} \sup_{c \in C} c'x$$
$$\text{s.t.} \quad a'x \geq b, (a, b) \in \bigcup_{t \in T} U_t,$$

to stochastic, fuzzy, interval, and parametric models. What the users expect from a robust optimization model is numerical tractability and existence of some optimistic counterpart (concept that we define in a precise way in Chap. 3), which is obtained via duality, both problems having the same optimal value. The adjective "tractable" means in (finite) linear and convex programming that there exists an equivalent problem for which there are known solution algorithms with worst-case running time polynomial in a properly defined input size [20]. Concerning uncertain LSIO, as this type of optimization problems are almost always hard, "tractable" means in this Brief that there exists some algorithm providing an ε-optimal solution in "reasonable time" for any $\varepsilon > 0$.

Decision makers less risk-averse may prefer a mixed model which combines, e.g., the robust approach w.r.t. the constraints with the probabilistic (fuzzy, interval, parametric) one w.r.t. the objective function. These mixed models are obtained from P_0 in (2.1) by interpreting c as a random (fuzzy, interval, parametric) vector. Let us mention the existence of a stream of works comparing probabilistic and robust models for certain types of finite uncertain optimization problems with random data, whose main aim consists of guaranteeing that, under suitable assumptions on the uncertain sets, any robust feasible solution satisfies the probabilistic constraints with high probability (see [20] and references therein).

2.1.5 The Parametric Approach

Parametric models are based on embedding the nominal problem $\overline{P}$ in (2.2), identi-fied with the triplet $\overline{\pi} = (\overline{c}, \overline{a}, \overline{b})$, into a suitable topological space of admissible perturbed problems, the so-called *space of parameters* Π. The topology on Π usually corresponds to some measure of the size of the admissible perturbations. When the perturbations are required to preserve the number n of decision variables and the index set T, there is a consensus about the convenience of measuring the distance between two parameters $\pi_1 = (c^1, a^1, b^1) \in \Pi$ and $\pi_2 = (a^2, b^2, c^2) \in \Pi$ by the box extended distance

$$d(\pi_1, \pi_2) := \max \left\{ \left\| c^1 - c^2 \right\|, \ \sup_{t \in T} \left\| (a_t^1, b_t^1) - (a_t^2, b_t^2) \right\| \right\}$$
$$= \max \left\{ \left\| c^1 - c^2 \right\|, \ d(\sigma_1, \sigma_2) \right\}, \tag{2.4}$$

with σ_1, σ_2 being the constraint systems of π_1, π_2, respectively, and where $\|\cdot\|$ is any norm in $\mathbb{R}^n$ and in $\mathbb{R}^{n+1}$. A relevant particular case is

$$d_\infty (\pi_1, \pi_2) = \max \left\{ \|c^1 - c^2\|_\infty, \sup_{t \in T} \|(a_t^1, b_t^1) - (a_t^2, b_t^2)\|_\infty \right\}, \qquad (2.5)$$

and in a similar way we shall define $d_2 (\pi_1, \pi_2)$.

These extended distances describe the topology of the uniform convergence. When all the data in the nominal problem $\overline{P}$, represented by the triplet $\overline{\pi} = (\overline{c}, \overline{a}, \overline{b})$, can be perturbed, the parameter space is $\Pi = (\mathbb{R}^n)^T \times \mathbb{R}^T \times \mathbb{R}^n$, in the general case, and $\Pi = \mathcal{C}(T)^n \times \mathcal{C}(T) \times \mathbb{R}^n$, in the continuous case. Observe that Π is a real linear space for the componentwise operations and it is not connected in the general case as the sets of parameters with bounded and unbounded constraints data sets are complementary open cones.

Qualitative stability analysis provides conditions under which sufficiently small perturbations of the nominal problem provoke small changes in the optimal value, the optimal set, and the feasible set. The *(primal) optimal value function* is the single-valued extended function $\vartheta : \Pi \to \overline{\mathbb{R}}$ such that $\vartheta(\pi)$ is the optimal value of P (i.e., $\vartheta(\pi) = v(P)$), whose desirable stability properties are the lower and upper semicontinuity. The optimal set and the feasible set mappings are set-valued mappings. The *(primal) feasible set mapping* $\mathcal{F} : \Pi \rightrightarrows \mathbb{R}^n$ associates with each $\pi \in \Pi$ the feasible set $\mathcal{F}(\pi)$ of P (the LSIO problem associated with π) while the *(primal) optimal set mapping* $\mathcal{S} : \Pi \rightrightarrows \mathbb{R}^n$ associates with each $\pi \in \Pi$ the optimal set $\mathcal{S}(\pi)$ of P. In a similar way we shall consider the *dual optimal value function* $\vartheta^D : \Pi \to \overline{\mathbb{R}}$, the *dual feasible set mapping* $\mathcal{F}^D : \Pi \rightrightarrows \mathbb{R}_+^{(T)}$ and the *dual optimal set mapping* $\mathcal{S}^D : \Pi \rightrightarrows \mathbb{R}_+^{(T)}$ assigning to $\pi \in \Pi$ the optimal value, the feasible set and the optimal set of D, respectively. Observe that ϑ (and also ϑ^D) is *positively homogeneous*, i.e., $\vartheta(\mu\pi) = \mu\vartheta(\pi)$ for any $\mu > 0$ and $\pi \in \Pi$.

At this point we must recall the basic continuity concepts for set-valued mappings (as $\mathcal{F}$ and $\mathcal{S}$). Consider two topological spaces Y and X (in our parametric framework, $X = \mathbb{R}^n$ is the space of decisions of the nominal problem while Y is some topological subspace of the space of parameters Π). Consider a set-valued mapping $\mathcal{M}$ between Y and X, i.e., $\mathcal{M} : Y \rightrightarrows X$, and $\overline{y} \in Y$ such that $\mathcal{M}(\overline{y}) \neq \emptyset$. Then :

- $\mathcal{M}$ is *(Berge-Kuratowski) lower semicontinuous* (lsc, in brief) at $\overline{y}$ if for each open set $V \subset X$ verifying $\mathcal{M}(\overline{y}) \cap V \neq \emptyset$, there exists $U \in \mathfrak{N}_{\overline{y}}$ such that $\mathcal{M}(y) \cap V \neq \emptyset$, for all $y \in U$.
- $\mathcal{M}$ is *(Berge-Kuratowski) upper semicontinuous* (usc, shortly) at $\overline{y}$ if for each open set $V \subset X$ verifying $\mathcal{M}(\overline{y}) \subset V$, there exists $U \in \mathfrak{N}_{\overline{y}}$ such that $\mathcal{M}(y) \subset V$, for all $y \in U$.
- If Y and X are pseudometric spaces, $\mathcal{M}$ is said to be *closed* at $\overline{y}$ if for all sequences $(y_r)_{r=1}^\infty \subset Y$ and $(x_r)_{r=1}^\infty \subset X$ satisfying $x_r \in \mathcal{M}(y_r)$ for all $r \in \mathbb{N}$, $\lim_{r \to \infty} y_r = \overline{y}$ and $\lim_{r \to \infty} x_r = \overline{x}$, one has $\overline{x} \in \mathcal{M}(\overline{y})$.

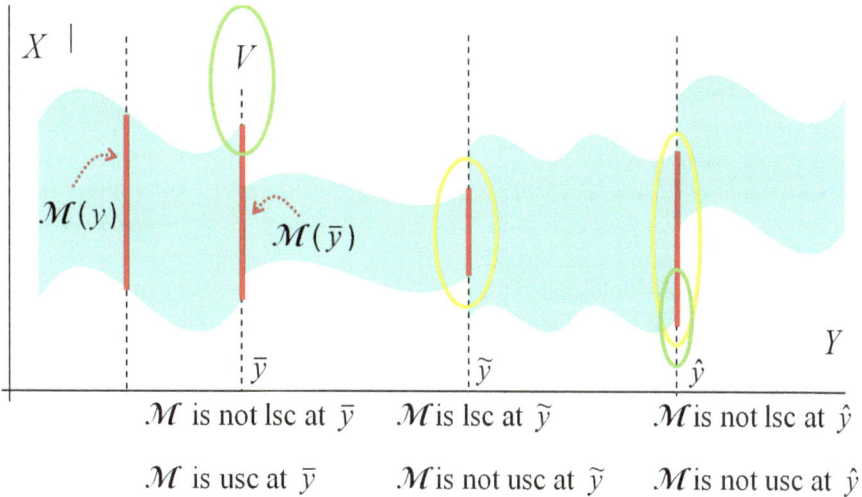

$\mathcal{M}$ is not lsc at $\overline{y}$ $\mathcal{M}$ is lsc at $\widetilde{y}$ $\mathcal{M}$ is not lsc at $\hat{y}$

$\mathcal{M}$ is usc at $\overline{y}$ $\mathcal{M}$ is not usc at $\widetilde{y}$ $\mathcal{M}$ is not usc at $\hat{y}$

Fig. 2.1 Lower and upper semicontinuity

Let us observe that if $\mathcal{M}$ is a single-valued mapping, lower and upper semicontinuity coincide with the ordinary notion of continuity of $\mathcal{M} : Y \rightarrow X$. Continuity also implies closedness, but the converse is not true (recall the function f in (2.3)).

The *graph* and the *domain* of $\mathcal{M}$ are

$$\text{gph } \mathcal{M} := \{(y, x) \in Y \times X : x \in \mathcal{M}(y)\}$$

and

$$\text{dom } \mathcal{M} := \{y \in Y : \mathcal{M}(y) \neq \emptyset\} = \text{Proj}_Y \text{ gph } \mathcal{M},$$

respectively.

Lower semicontinuity precludes the image set $\mathcal{M}(y)$ to shrink drastically for y close to $\overline{y}$, whereas upper semicontinuity avoids the opposite situation, in other words, that the image sets explode in size around $\overline{y}$. Closedness of $\mathcal{M}$ at $\overline{y}$ means that gph $\mathcal{M}$ contains the limits of its sequences whose images by Proj_Y gph $\mathcal{M}$ converge to $\overline{y}$. In Fig. 2.1 the graph of a set-valued mapping $\mathcal{M} : \mathbb{R} \rightrightarrows \mathbb{R}$ is the blue shaded area, and we show three different situations at the points $\overline{y}$ (usc, not lsc), $\widetilde{y}$ (lsc, not usc), and $\hat{y}$ (not lsc, not usc). Moreover, $\mathcal{M}$ is non-closed at $\widetilde{y}$ and $\hat{y}$.

Closedness of $\mathcal{M}$ at every $y \in \text{dom } \mathcal{M}$ is equivalent to the closedness of gph $\mathcal{M}$ in the product space $Y \times X$. Closedness and upper semicontinuity compete to be the counterpart concept of the lower semicontinuity.

It makes sense to consider that the feasible set mapping $\mathcal{F}$ is also defined on the space Θ of systems associated with the parameters $\sigma = (a, b)$ (these systems are also called *semi-infinite* and appear in different fields). In this way we may use indistinctly $\mathcal{F}(\pi)$ or $\mathcal{F}(\sigma)$. Similarly for the dual feasible mapping, i.e., $\mathcal{F}^D(\pi)$ is the same thing that $\mathcal{F}^D(\sigma_D)$ where $\sigma_D := \{\sum_{t \in T} \lambda_t a_t = c\}$. Observe that $\mathcal{F}$ is closed for free. Indeed, let $\overline{\sigma} = \left(\overline{a}, \overline{b}\right) \in \Theta$, and two sequences $(\sigma_r)_{r=1}^{\infty} \subset \Theta$ and $(x_r)_{r=1}^{\infty} \subset \mathbb{R}^n$ satisfying $x_r \in \mathcal{F}(\pi_r)$ for all $r \in \mathbb{N}$, $\lim_{r \to \infty} \sigma_r = \overline{\sigma}$ and $\lim_{r \to \infty} x_r = \overline{x}$. Let $\sigma_r = \{\langle a^r(t), x \rangle \geq b^r(t), t \in T\}$, $r \in \mathbb{N}$. Fixed $t \in T$, taking limits as $r \to \infty$ in $\langle a^r(t), x_r \rangle \geq b^r(t)$, one has $\langle \overline{a}(t), \overline{x} \rangle \geq \overline{b}(t)$. Hence, $\overline{x} \in \mathcal{F}(\overline{\sigma})$ and so $\mathcal{F}$ is closed at $\overline{\sigma}$. We conclude that, in the framework of the stability of the feasible set, closedness is too weak while upper semicontinuity is too strong (it hardly holds when the image is non-compact). Concerning $\mathcal{S}$, we will see in Sect. 5.1 that its closedness coincides essentially with the lsc property of $\mathcal{F}$. The stability of $\mathcal{F}^D : \Pi \rightrightarrows \mathbb{R}_+^{(T)}$ has only been analyzed by assuming that $\mathbb{R}^{(T)}$ is equipped with the $\|\cdot\|_{\infty}$ and $\|\cdot\|_1$ norms [105] while the stability of $\mathcal{S}^D$ has not been considered yet.

Quantitative stability analysis deals with error bounds on distances in the decision and the parameter spaces. A fundamental formula obtained by Hoffman in 1952 [135] provides an error bound on the distance from any point of $\mathbb{R}^n$ to the solution set of a linear system in terms of the most violated constraint. This formula turned out to be related to the complexity of numerical methods in LO. For this reason, many extensions and variations of this result have been proposed for different types of systems, including those arising in LSIO. The parametric models also allow to compute the *distance to ill-posedness* relative to some property enjoyed by the nominal problem, i.e., the minimum size of those perturbations of $\overline{P}$ which provide perturbed problems not enjoying such a property.

Sensitivity analysis provides estimations of the impact of a given perturbation on the optimal value. Most of the existing literature on sensitivity analysis in LO has focused on determining linearity (or affinity) regions of the optimal value function under perturbations of the vector cost $\overline{c}$, or perturbations of the right-hand side (RHS) vector $\overline{b}$, or both (it is difficult to study the effect on the optimal value of perturbations of the left-hand side $\overline{a}$ even in LO). Sensitivity analysis in LO can be approached from three different perspectives. The classical approach is based on the use of optimal basis (the available information when the simplex method attains an optimal extreme point of the polyhedral feasible set). This approach cannot be extended to LSIO because the number of active constraints at an extreme point of the feasible set is seldom more than one (in particular, the smooth convex bodies as Euclidean balls and ellipsoids have a unique supporting hyperplane at every boundary point). The duality approach provides conditions for the affinity on segments or half-lines of the optimal value function for non-simultaneous perturbations of costs and RHS. The third approach to sensitivity analysis in LO exploits the optimal partitions computed by the interior point method when optimality is achieved, allowing to treat simultaneous perturbations of both, costs and RHS.

Example 2.1.1. Consider the LSIO problem, say $\overline{P}$, of Example 1.1.1(c) and assume that the RHS of the subsystem corresponding to indexes $t \in \left[0, \frac{\pi}{2}\right]$ can be perturbed by a common parameter $y \in \mathbb{R}$ (to be precise, the parameter space is the set of triplets $(\overline{c}, \overline{a}, b)$, where $b_t = -y$ for all $t \in \left[0, \frac{\pi}{2}\right]$, and it can be identified with $\mathbb{R}$). Then we have to consider the parametric LSIO problem

$$P(y) : \inf_{x \in \mathbb{R}^2} x_1$$
$$\text{s.t.} \quad -(\cos t) x_1 - (\sin t) x_2 \geq -y, \ t \in \left[0, \tfrac{\pi}{2}\right],$$
$$x_1 \geq 0 \ (t = 2), \ x_2 \geq 0 \ (t = 3).$$

Observe that $P(1) \equiv \overline{P}$. Here $\mathcal{F}, \mathcal{S} : \mathbb{R} \rightrightarrows \mathbb{R}^2$ are $\mathcal{F}(y) = \{x \in \mathbb{R}^2_+ : \|x\|_2 \leq y\}$ and $\mathcal{S}(y) = \mathcal{F}(y) \cap (\{0\} \times \mathbb{R})$ while $\vartheta(y) = 0$ if $y \in \mathbb{R}_+$ and $\vartheta(y) = +\infty$ otherwise. So, we have $\operatorname{dom} \mathcal{F} = \operatorname{dom} \mathcal{S} = \operatorname{dom} \vartheta = \mathbb{R}_+$, $\operatorname{gph} \vartheta = \mathbb{R}_+ \times \{0\}$, $\operatorname{gph} \mathcal{F} = \{(y, x_1, x_2) \in \mathbb{R}^3_+ : \|x\|_2 \leq y\}$ (see Fig. 2.2) and $\operatorname{gph} \mathcal{S} = \{(y, 0, x_2) \in \mathbb{R}^3_+ : 0 \leq x_2 \leq y\}$ (see Fig. 2.3).

Since epi ϑ is closed while epi $(-\vartheta)$ is not, ϑ is lsc but not usc (the latter property fails at $y = 0$). Concerning $\mathcal{F}$ and $\mathcal{S}$, the situation is exactly the opposite: they are usc but not lsc. In fact, the latter property fails at $y = 0$ as $\mathcal{S}(0) = \mathcal{F}(0) = \{0_2\}$ while $\mathcal{S}(y) = \mathcal{F}(y) = \emptyset$ for all $y < 0$. Moreover, $\mathcal{S}$ is closed due to the closedness of $\operatorname{gph} \mathcal{S}$ (recall that $\mathcal{F}$ is always closed).

Concerning the stability of $\overline{P}$, since the three mappings are lsc and usc at $\overline{y} = 1$, and $\mathcal{F}$ and $\mathcal{S}$ are also closed at that parameter, we conclude that the nominal parameter is stable in all senses defined up to now (other stability concepts will be introduced later).

Observe that the distance from $\overline{P}$ (or the corresponding parameter $\overline{y} = 1$) to inconsistency is 1 because $\mathcal{F}(y) \neq \emptyset$ for all $y \in \mathbb{R}$ such that $d_\infty(y, 1) = |y - 1| \leq 1$ while $\mathcal{F}\left(-\frac{1}{r}\right) = \emptyset$ for all $r \in \mathbb{N}$, with $d_\infty\left(-\frac{1}{r}, 1\right) = 1 + \frac{1}{r} \to 1$. The problem $P(0)$ is ill-posed in the consistency sense as any neighborhood of 0 contains both consistent and inconsistent problems. Since $\operatorname{dom} \mathcal{S} = \operatorname{dom} \vartheta$, we can replace "consistency" by "solvability" in the last two paragraphs.

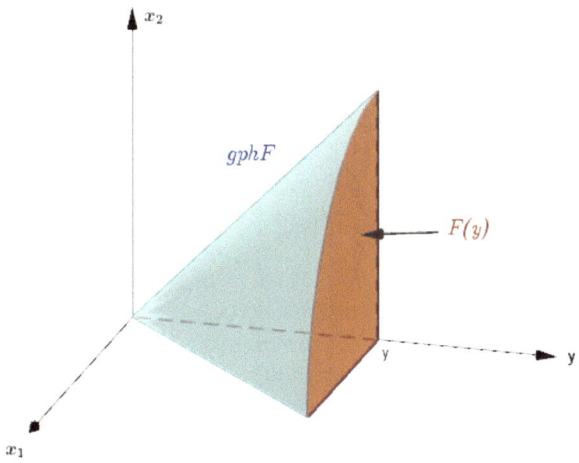

Fig. 2.2 Graph of the feasible set mapping

Fig. 2.3 Graph of the
optimal set mapping

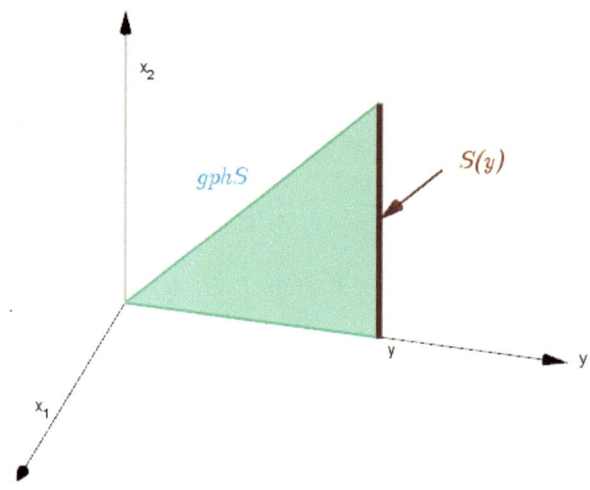

Finally, notice that the sensitivity analysis is trivial in this example since ϑ is constant on $\mathbb{R}_+$ (i.e., changes smaller than 1 in the nominal problem have no effect on ϑ).

The parametric model most frequently encountered in the recent literature on stability in LSIO considers perturbations of all data. One reason for this is that the characterizations of different qualitative stability properties in this model become sufficient conditions for the remaining models and sometimes these conditions are also necessary. Analogously, the formulae providing the distance to ill-posedness are at least upper bounds in other models whereas the error bound are still valid (although they could be improved). Sometimes it is more difficult to study the stability under perturbations of part of the data (usually the LHS $\overline{a}$) than under the perturbations of all data while the continuous models allow to use useful topological and analytical tools as the Urisohn's lemma or Robinson–Ursescu theorem (provided that gph $\mathcal{F}$ is convex, as it is whenever $\overline{a}$ is fixed). Few sensitivity analysis results have been published on LO under perturbations of data including $\overline{a}$, so that the reader only can expect sensitivity analysis results on LSIO under perturbations of the remaining data; thus, Sects. 4.1, 4.2, and 4.3 consider perturbations of $\overline{c}$, $\overline{b}$, and the couple $(\overline{b}, \overline{c})$, respectively. Simultaneous perturbations of $(\overline{c}, \overline{a}, \overline{b})$ are considered in Sects. 5.1, 5.3, 6.2.1, and 6.3.1, separate perturbations of $\overline{a}, \overline{b}$, and $\overline{c}$ in Sect. 5.2, perturbations of $\overline{b}$ and $(\overline{a}, \overline{b})$ in Sect. 6.2.2, and perturbations of $(\overline{c}, \overline{b})$ in Sect. 6.3.2.

The convenience and viability of the alternative models sketched in this section for a specific uncertain LSIO problem depend on the nature of its data, the attitude of the decision maker toward risk, the tractability of the auxiliary problems to be solved, and the availability of hardware and software facilities. In conclusion, instead of claiming the superiority of one of the above models on the others, the authors agree with the famous statement of the statistician George Box: "all models are wrong, but some are useful" [28].

Remark 2.1.1 (Antecedents).
The Stochastic Approach: The existing literature on probabilistic models for uncertain semi-infinite problems [67–69, 211] does not include contributions on LSIO.

The Fuzzy Approach: The first works on fuzzy optimization were published in the 1970s [13, 216]. In particular, [235] showed the way to reduce fuzzy LO problems with linear membership functions to ordinary LO problems. There exists a wide literature on fuzzy optimization via ranking functions (see, e.g., [76, 79, 171, 228], and the review paper [32]). The survey paper [170] reformulates and solves fuzzy LO problems with crisp objective function as LSIO ones, and the same methodology could be applied to LSIO problems with fuzzy constraints. To the authors knowledge, no work has been published on fuzzy LSIO. Other connections between fuzzy optimization and LSIO have been established in [138, 195], and [227], where convex fuzzy optimization problems, semi-infinite fuzzy systems with concave membership functions (whose unique illustrative example is linear), and optimal design of alumina ceramics with fuzzy data, respectively, have been reformulated as LSIO problems.

The Interval Approach: Bhattacharjee et al. [24] review the interval approach in nonlinear semi-infinite optimization without considering the particular case of LSIO. The interval approach has been used to tackle uncertain LSIO problems arising in environmental engineering in a rather empirical way [125, 126, 129, 130, 142, 173, 237]. In the more complex interval model of [173] the decision variables are replaced with decision intervals.

The Robust Approach: LO problems with uncertain constraints have been treated in a robust way in [214] under the name of inexact linear programming, and their relationship with LSIO was explored in [219], whose authors consider uncertain constraints (with coefficient vectors (a_t, b_t) typically ranging on some ball for a weighted supremum norm). The antecedents of robust LSIO are reviewed in Chap. 3.

The Parametric Approach: The extended distance defined in (2.5) was introduced in the seminal paper [123]. The first results on the stability of $\mathcal{F}$, $\mathcal{S}$, and ϑ were published in the 1980s and dealt with continuous LSIO problems. The papers [92] and [176] have recently reviewed the existing vast literature on the parametric model associated with the pseudometric d defined in (2.4) for LSIO, LIO, and CIO. More details can be found in the remarks to Sects. 4–6.

2.2 Modeling Uncertain Portfolio Selection

Assume that C euros are to be invested in a portfolio comprised of n assets (shares, stocks, securities). Let r_i be the return per 1 euro invested in asset $i \in \{1, \ldots, n\}$ during a period of time (e.g., 1 month or 1 year). Since these returns are not known in advance, $r = (r_1, \ldots, r_n)$ is an uncertain vector. The decision variables are the amount of euros to be invested in the ith asset, denoted by $x_i, i = 1, \ldots, n$. Any feasible portfolio $x = (x_1, \ldots, x_n)$ satisfies $x_i \geq 0$, $i = 1, \ldots, n$, $\sum_{i=1}^{n} x_i = C$, and other linear constraints, at least one of them being uncertain. In this section we discuss portfolio models which are robust w.r.t. the constraints; one of these models is mixed whenever r is interpreted as a random (fuzzy, interval, parametric) vector while it is a pure robust model when the task consists of maximizing the worst possible return of the portfolio. Recalling (2.1), we consider an LSIO uncertain problem of the form

$$P_0 : \max_{x \in \mathbb{R}^n} r'x := \sum_{i=1}^{n} r_i x_i \text{ s.t. } \overline{a}_t'x \geq \overline{b}_t, t \in T,$$

where the uncertainty exclusively affects the objective function. We assume the feasibility of P_0, i.e., the nonemptiness of the feasible set $F \subset \{x \in \mathbb{R}_+^n : \sum_{i=1}^{n} x_i = C\}$.

We will get a *probabilistic model* by interpreting each rate of return r_i as a random variable with expected value $\mathbb{E}[r_i]$, $1, \ldots, n$. In the classical model of stochastic optimization for the portfolio problem, whose history can be traced back to the 1950s, the decision maker solves

$$P_1 : \max_{x \in \mathbb{R}^n} \mathbb{E}[r'x] := \sum_{i=1}^{n} \mathbb{E}[r_i] x_i \text{ s.t. } \overline{a}_t'x \geq \overline{b}_t, t \in T,$$

where the maximum is attained due to the compactness of F. This model is very simple but absolutely unrealistic because it does not take risk into account. Now we assume that the investor wants to get a return less than $\eta \in I \subset \mathbb{R}$ (I is an interval containing any conceivable return) from a feasible portfolio x with a probability $\Pr\{r'x \leq \eta\}$ not greater than some $p_\eta \in [0, 1]$. The function $\eta \mapsto p_\eta$ captures the attitude of the decision maker toward risk and each inequality $\Pr\{r'x \leq \eta\} \leq p_\eta$ is called a *Value-at-Risk* (or a stochastic dominance) *constraint*. Thus we get the following probabilistic problem:

$$P_2 : \max_{x \in \mathbb{R}^n} \mathbb{E}[r'x]$$
$$\text{s.t.} \quad \Pr\{r'x \leq \eta\} \leq p_\eta, \ \eta \in I, \qquad (2.6)$$
$$\overline{a}_t'x \geq \overline{b}_t, t \in T.$$

To the authors knowledge, the few available numerical methods for stochastic dominance constrained programs [137, 141] do not allow to solve the nonlinear semi-infinite problem P_2 except in particular cases. In fact, P_2 is a hard problem even in the simplest case that $|I| = 1$ and $|T| < \infty$ (as in the classical chance constrained portfolio model), unless r has a multivariate normal distribution.

We shall get a *robust optimization model* by assuming that r ranges on a given bounded set $R \subset \mathbb{R}^n$. A pessimistic decision maker should maximize the worst possible return of the portfolio x, i.e., the number $\inf_{r \in R} r'x$. Thus, the robust counterpart of P_0 is the maxmin problem

$$P_3 : \sup_{x \in \mathbb{R}^n} \inf_{r \in R} r'x \text{ s.t. } \overline{a}'_t x \geq \overline{b}_t, t \in T. \tag{2.7}$$

Observing that $\inf_{r \in R} r'x$ is the greatest lower bound of the set $\{r'x : r \in R\}$, the nonlinear semi-infinite program P_3 turns out to be equivalent to the LSIO problem

$$P_4 : \sup_{(x,y) \in \mathbb{R}^{n+1}} y$$
$$\text{s.t.} \qquad r'x - y \geq 0, \ r \in R, \tag{2.8}$$
$$\overline{a}'_t x \geq \overline{b}_t, \ t \in T.$$

If $\sigma = \{\overline{a}'_t x \geq \overline{b}_t, t \in T\}$ is continuous and R is a compact convex subset of $\mathbb{R}^n$, then P_4 is a continuous LSIO problems, so that it can be solved by means of the discretization methods mentioned in Sect. 1.3 except the interior point constraint generation algorithm (because the feasible set of P_4 has an empty interior), although grid discretization is not efficient whenever $\dim R \geq 3$ (frequently, $\dim R = n$). When T is finite it is preferable to use the interior point method of [215], exploiting the affinity of the function $r \mapsto r'x - y$ for any couple $(x, y) \in \mathbb{R}^{n+1}$. Numerical experiments with P_4, with T finite, are reported in the latter paper.

Next we propose an *interval optimization model* by assuming that the uncertain set is the box $R = \prod_{i=1}^{n} [\underline{r}_i, \overline{r}_i]$, with $\underline{r}_i < \overline{r}_i$, $i = 1, \ldots, n$. We must determine the range of the optimal value for all instances of the uncertain problem P_0, i.e., we have to solve a pessimistic and an optimistic counterparts of P_0, but exploiting the special structure of R. Since $\{\underline{r}_i, \overline{r}_i\}^n$ is the set of extreme points of R, by a convexity argument, the constraint subsystem $\{r'x - y \geq 0, r \in R\}$ is equivalent to $y \leq \sum_{i=1}^{n} \min\{\underline{r}_i x_i, \overline{r}_i x_i\}$. So, introducing an auxiliary variable $z \in \mathbb{R}^n$, P_4 can be reformulated as

$$P_5 : \sup_{(x,y,z) \in \mathbb{R}^{n+1+n}} y$$
$$\text{s.t.} \qquad \sum_{i=1}^{n} z_i - y \geq 0,$$
$$\underline{r}_i x_i - z_i \geq 0, \ i = 1, \ldots, n,$$
$$\overline{r}_i x_i - z_i \geq 0, \ i = 1, \ldots, n,$$
$$\overline{a}'_t x \geq \overline{b}_t, \ t \in T.$$

The choice of a suitable LSIO numerical method to solve P_5 is conditioned to the properties of the system $\{\overline{a}'_t x \geq \overline{b}_t, t \in T\}$. An optimistic decision maker expects to get a return $\max_{r \in R} r'x$ from a feasible portfolio x. Thus she/he must solve the quadratic semi-infinite optimization problem

$$P_6 : \max_{(x,r) \in \mathbb{R}^{2n}} r'x$$
$$\text{s.t.} \qquad \underline{r}_i \leq r_i \leq \overline{r}_i, \ i = 1, \ldots, n,$$
$$\overline{a}'_t x \geq \overline{b}_t, \ t \in T,$$

whose difficulty lies in the lack of convexity of the objective function.

Finally, we construct a *parametric model* from a given nominal problem

$$\overline{P} : \max_{x \in \mathbb{R}^n} \overline{r}'x \text{ s.t. } \overline{a}_t'x \geq \overline{b}_t, t \in T.$$

The space of parameters is $\Pi_1 := \{\overline{a}\} \times \{\overline{b}\} \times \mathbb{R}^n$, that we can identify with $\mathbb{R}^n$, equipped with the supremum distance d_∞. Obviously, $\mathcal{F}|_{\Pi_1} : \mathbb{R}^n \rightrightarrows \mathbb{R}^n$ is constant, while $\mathcal{S}|_{\Pi_1} : \mathbb{R}^n \rightrightarrows \mathbb{R}^n$ and $\vartheta|_{\Pi_1} : \mathbb{R}^n \to \mathbb{R}$ due to the compactness of F. We shall prove that $\mathcal{S}|_{\Pi_1}$ is closed and usc at $\overline{r}$ while $\vartheta|_{\Pi_1}$ is continuous at $\overline{r}$.

Let $(r_k)_{k=1}^\infty \subset \mathbb{R}^n$ and $(x^k)_{k=1}^\infty \subset \mathbb{R}^n$ be such that $x^k \in \mathcal{S}(r_k)$ for all $k \in \mathbb{N}$, $r_k \to \overline{r}$ and $x^k \to \overline{x}$. Given $x \in F$, we have

$$r_k'x^k \geq r_k'x \text{ for all } k \in \mathbb{N}. \tag{2.9}$$

Taking limits in (2.9) as $k \to \infty$, we get $\overline{r}'\overline{x} \geq \overline{r}'x$, so that $\overline{x} \in \mathcal{S}(\overline{r})$. This shows that $\mathcal{S}$ is closed at $\overline{r}$. Moreover, $\mathcal{S}$ is *equibounded* at $\overline{r}$ in the sense that there exists $U \in \mathfrak{N}_{\overline{r}}$ such that $\cup\{\mathcal{S}(r) : r \in U\}$ is bounded (take, e.g., $U = \mathbb{R}^n$). Since any equibounded set-valued mapping which is closed at a given parameter is usc at that parameter, $\mathcal{S}$ turns out to be usc at $\overline{r}$.

The following example shows that $\mathcal{S}|_{\Pi_1}$ is not necessarily lsc at $\overline{r}$: if $F = [0,1]^2$ and $\overline{r} = (0,1)$, then $\mathcal{S}(\overline{r}) = [0,1] \times \{1\}$ while $\mathcal{S}(r) \subset \{(0,1),(1,1)\}$ for $r \notin \text{cone}\{\overline{r}\}$ sufficiently close to $\overline{r}$, so that $\mathcal{S}$ shrinks abruptly close to $\overline{r}$.

Concerning the optimal value function, since

$$\vartheta(r) = \max_{x \in F} r'x,$$

ϑ coincides with the *support function* σ_F of the compact convex set F (i.e., $\sigma_F(r) = \sup_{x \in F} x'r$), ϑ is a finite-valued positively homogeneous convex function, which implies $\vartheta \in \mathcal{C}(\mathbb{R}^n)$. Thus, ϑ is continuous at $\overline{r}$.

Remark 2.2.1 (Antecedents). Examples of P_2, in (2.6), with real data from US stock markets, with T finite, are discussed and solved in [68] and [69]. Alternative models are obtained by replacing the Value-at-Risk constraints in P_3, in (2.7), by convex approximations involving moments instead of probabilities. The tightest approximations are provided by a risk measure introduced by Ben-Tal and Teboulle [17] and later popularized by Rockafellar and Uryasev [206] under the name of conditional Value-at-Risk measure. This way P_2 is replaced by a suitable convex semi-infinite optimization (CSIO) problem which admits an equivalent LSIO reformulation under mild assumptions [71].

There exists a wide literature on fuzzy models for the portfolio problem (see, e.g., the recent surveys [225, 238]).

The returns $r_i's$ are interpreted as trapezoidal fuzzy numbers in [168, 170], where T is finite. In [224] the $r_i's$ are interpreted as LR-fuzzy numbers of different forms. This type of fuzzy numbers were defined by Dubois and Prade [77] in an axiomatic way, enumerating the properties of two auxiliary functions, denoted by L and R, which are required to be even, decreasing, usc on their supports, and take value 1 at 0. The first model proposed in [224] consists of minimizing certain measure of the risk of achieving a return that is less than the return η of a given riskless asset (i.e., $\sum_{i=1}^{n} x_i r_i \leq \eta$) subject to $a'_t x \geq b_t, t \in T$. The second model replaces this unique fuzzy constraint with infinitely many linear constraints giving rise to a certain LSIO problem. Numerical experiments with the latter model and T finite have been reported in [224], taking real data from the Spanish stock market and making use of the simplex-like method in [169].

Remark 2.2.2 (An Open Problem in Uncertain LSIO). There is an absolute lack of software implementations and applications of post-optimality techniques to solve real-world LSIO problems.

Chapter 3
Robust Linear Semi-infinite Optimization

For many finite optimization problems, numerical methods can be compared from the complexity point of view, i.e., computing upper bounds on the number of iterations, arithmetic operations, etc., necessary to get an optimal solution, or an ε-optimal solution, in terms of the size of the problem. This methodology can hardly be applied in LSIO because it is not evident how to define the size of the triplet (a, b, c) representing the data of a problem like (1.1) despite the seminal results on the complexity of the interior point constraint generation algorithm in [182, 192]. On the other hand, the robust counterpart of an uncertain LSIO problem seldom enjoys the strong assumptions which are necessary to apply reduction or feasible point methods. For this reason we identify, in this framework, tractability of a given LSIO problem with satisfaction of the conditions guaranteeing the viability of the discretization methods, namely:

- Continuity of the problem (main ingredient of any convergence proof).
- The density assumption and low dimension of the index set (for grid discretization algorithms).
- The boundedness of the optimal set (for central cutting plane algorithms).
- The full dimension and boundedness of the feasible set (for the interior point constraint generation algorithm).

In this chapter we consider given an uncertain LSIO problem

$$P_0 : \inf_{x \in \mathbb{R}^n} c'x$$
$$\text{s.t.} \quad a_t'x \geq b_t, \ t \in T,$$

where the uncertain data may be either the constraints or the objective function, with $c = \overline{c} \in \mathbb{R}^n$ (a fixed vector) in the first case and $(a, b) = \left(\overline{a}, \overline{b}\right) \in (\mathbb{R}^n)^T \times \mathbb{R}^T$ (a fixed function) in the second one. The treatment of problems where all the data are uncertain is a combination of both models. In each case, we associate with P_0 a LSIO problem P_R called *robust* (or *pessimistic*) *counterpart* whose corresponding cost vector, index set, feasible set, optimal solution set, first moment cone, and

M.A. Goberna and M.A. López, *Post-Optimal Analysis in Linear Semi-Infinite Optimization*, SpringerBriefs in Optimization, DOI 10.1007/978-1-4899-8044-1_3,

characteristic cone are denoted by c_R, T_R, F_R, S_R, M_R, and K_R, respectively. The Haar dual problem of P_R is denoted by D_R. We also associate with P_0 a *robust dual problem* (also called *optimistic counterpart*) D^R such that the weak duality is always satisfied while the strong duality holds under mild conditions. We also examine the consistency of P_R and its numerical tractability.

3.1   Uncertain Constraints

The LSIO problem P_0 in the face of data uncertainty in the constraints can be captured by a parameterized LSIO problem of the form

$$P^u : \inf_{x \in \mathbb{R}^n} \bar{c}'x$$
$$\text{s.t.} \quad v_t'x \geq w_t, \forall t \in T,$$

where $\bar{c} \in \mathbb{R}^n$ and $u = (v, w) : T \to \mathbb{R}^n \times \mathbb{R}$ represents a *selection* of a given *uncertain set-valued mapping* $\mathcal{U} : T \rightrightarrows \mathbb{R}^{n+1}$ (in short, $u \in \mathcal{U}$). Let $U_t := \mathcal{U}(t) \subset \mathbb{R}^{n+1}$ for all $t \in T$. Hence, in this robust model, the uncertainty set is the *graph* of $\mathcal{U}$, that is, $\text{gph}\,\mathcal{U} = \{(t, u_t) : u_t \in U_t, t \in T\}$.

A robust decision maker facing uncertainty in the constraints intends to guarantee the feasibility of her/his decisions, so that the *robust counterpart* of the parametric problem $(P^u)_{u \in \mathcal{U}}$ is the deterministic problem

$$P_R : \inf_{x \in \mathbb{R}^n} \bar{c}'x$$
$$\text{s.t.} \quad v_t'x \geq w_t, u_t = (v_t, w_t) \in U_t, t \in T, \tag{3.1}$$

where the uncertain constraints are enforced for every possible value of the data within the prescribed uncertainty set $\text{gph}\,\mathcal{U}$. Denoting $T_R := \bigcup_{t \in T} U_t$, we can also write

$$P_R : \inf_{x \in \mathbb{R}^n} \bar{c}'x$$
$$\text{s.t.} \quad v'x \geq w, (v, w) \in T_R. \tag{3.2}$$

So the feasible set, the characteristic cone and the first moment cones of P_R are $F_R = \{v'x \geq w, (v, w) \in T_R\}$,

$$K_R = \text{cone}\,\{T_R \cup \{(0_n, -1)\}\} \quad \text{and} \quad M_R = \text{cone}\,\text{Proj}_{\mathbb{R}^n}(T_R), \tag{3.3}$$

respectively.

Example 3.1.1. For illustration purposes, let us revisit Example 1.1.1 with $\bar{c} = (-1, -1)$, where one considers perturbations, $v_1(t)$ and $v_2(t)$, of the nominal LHS coefficients of $t \in \left[0, \frac{\pi}{2}\right]$, $-\cos t$ and $-\sin t$, such that $|v_1(t) + \cos t| \leq \alpha$ and $|v_2(t) + \sin t| \leq \alpha$, with $0 < \alpha < 1$ (α can be interpreted as an upper bound for the rounding errors caused by the subroutines providing approximate values of the LHS coefficients corresponding to indices $t \in [0, \frac{\pi}{2})$. In terms of the robust model considered in this section, the index set is $T = [0, \frac{\pi}{2}] \cup \{2, 3\}$, the uncertain set-valued mapping $\mathcal{U} : T \rightrightarrows \mathbb{R}^3$ is

$$U_t = \begin{cases} [-\cos t - \alpha, -\cos t + \alpha] \times [-\sin t - \alpha, -\sin t + \alpha] \times \{-1\}, \ t \in [0, \frac{\pi}{2}], \\ \{(1, 0, 0)\}, & t = 2, \\ \{(0, 1, 0)\}, & t = 3, \end{cases}$$

and the robust counterpart of P_0 is

$$P_R : \inf_{x \in \mathbb{R}^n} -x_1 - x_2 \tag{3.4}$$
$$\text{s.t.} \quad v_t' x \ge w_t, \ (t, (v_t, w_t)) \in \text{gph}\,\mathcal{U}.$$

If $0 < \gamma < \frac{1}{\sqrt{2}+2\alpha}$ and $(t, (v_t, w_t)) \in \text{gph}\,\mathcal{U}$, one has

$$\langle v_t, (\gamma, \gamma)\rangle = \gamma \left(v_1(t) + v_2(t)\right) \ge -\gamma \left(\cos t + \sin t + 2\alpha\right) \ge -\gamma \left(\sqrt{2}+2\alpha\right) > -1,$$

so that P_R satisfies SCQ. Then, since $\text{gph}\,\mathcal{U}$ is compact, P_R satisfies the FMCQ too and so the characteristic cone K_R of P_R is closed.

A more realistic choice of $\mathcal{U}$ for P_0 derives from the fact that the trigonometric functions are computationally approached via Taylor's formula. So, we could take U_2 and U_3 as above and

$$U_t = \left[-1 + \frac{t^2}{2} - \frac{t^4}{24}, -1 + \frac{t^2}{2}\right] \times \left[-t, -t + \frac{t^3}{6}\right] \times \{-1\}, \text{ for all } t \in \left[0, \frac{\pi}{2}\right].$$

Now, for each fixed selection $u = (v, w) \in \mathcal{U}$, the dual of P_0 is the uncertain optimization problem

$$D^u : \sup_{\lambda \in \mathbb{R}_+^{(T)}} \left\{\sum_{t \in T} \lambda_t w_t : \sum_{t \in T} \lambda_t v_t = \overline{c}\right\}.$$

The optimistic counterpart of $(D^u)_{u \in \mathcal{U}}$ is given by

$$D^R : \sup_{\substack{u=(v,w)\in\mathcal{U} \\ \lambda \in \mathbb{R}_+^{(T)}}} \left\{\sum_{t \in T} \lambda_t w_t : \sum_{t \in T} \lambda_t v_t = \overline{c}\right\}.$$

By construction,

$$v(D^R) \le v(P_R). \tag{3.5}$$

We say that *robust duality* holds for P_0 whenever (3.5) holds with equality and D^R is solvable, i.e.,

$$\inf_{x \in F_R} \langle \overline{c}, x \rangle = \max_{\substack{u=(v,w) \in \mathcal{U} \\ \lambda \in \mathbb{R}_+^{(T)}}} \left\{ \sum_{t \in T} \lambda_t w_t : \sum_{t \in T} \lambda_t v_t = \overline{c} \right\} \tag{3.6}$$

whenever the first member of (3.6) is finite. This is the "primal worst value" while the second member is the "dual best value," so that robust duality means "primal worst equals dual best" with dual attainment.

We associate with P_R the *robust moment cone*

$$M^R := \bigcup_{u=(v,w) \in \mathcal{U}} \text{cone}\{(v_t, w_t), t \in T; (0_n, -1)\}.$$

Observe that M^R is generally neither convex nor closed and it is related to K_R by the equation $K_R = \text{conv } M^R$.

Now we are in a position to compare the optimistic counterpart D^R with the Haar dual D_R of the pessimistic counterpart of P_0 :

$$D_R : \sup_{\lambda \in \mathbb{R}_+^{(\text{gph}\,\mathcal{U})}} \sum_{(t,u_t) \in \text{gph}\,\mathcal{U}} \lambda_{(t,u_t)} b_{(t,u_t)}$$
$$\text{s.t.} \qquad \sum_{(t,u_t) \in \text{gph}\,\mathcal{U}} \lambda_{(t,u_t)} a_{(t,u_t)} = \overline{c}.$$

Recall (see Sect. 1.2) that D_R is equivalent to $\sup_{y \in \mathbb{R}} \{y : (\overline{c}, y) \in K_R\}$ in the sense that both problems have the same optimal value and are simultaneously solvable or not. By Theorem 1.2.1, $v(P_R) = v(D_R)$ whenever K_R is closed. If M^R is closed and convex, then $K_R = \text{conv } M^R = M^R$ is closed too and so

$$\begin{aligned} v(P_R) &= v(D_R) \\ &= \max \{y : (\overline{c}, y) \in K_R\} \\ &= \max \{y : (\overline{c}, y) \in M^R\} \\ &= v(D^R). \end{aligned}$$

We have thus proved the following robust duality theorem.

Theorem 3.1.1 (Global Robust Duality). *Let* $F_R \neq \emptyset$. *Then, robust duality holds for* P_0 *whenever the robust moment cone* M^R *is closed and convex.*

The following result is a local version of the robust duality theorem.

Theorem 3.1.2 (Local Robust Duality). *Suppose that the LFMCQ holds at* $\overline{x} \in F_R$ *and* U_t *is convex for all* $t \in T$. *Then, there exists a feasible solution* $\left((\overline{v}, \overline{w}), \overline{\lambda} \right)$ *for* D^R *such that* $\overline{c}'\overline{x} = \sum_{t \in T} \overline{\lambda}_t \overline{w}_t$.

In most real situations the uncertain set-valued mapping $\mathcal{U}$ takes the form

$$U_t := \left(\overline{a}_t, \overline{b}_t \right) + \alpha_t Z, \ t \in T, \tag{3.7}$$

where $\left(\overline{a},\overline{b}\right) \in \Theta := (\mathbb{R}^n)^T \times \mathbb{R}^T$, $\alpha \in \mathbb{R}_+^T$ and $Z \subset \mathbb{R}^{n+1}$ is a compact set such that $0_{n+1} \in Z$. Denote by $\overline{F}$ the feasible set of the nominal system $\left\{\overline{a}_t'x \geq \overline{b}_t, t \in T\right\}$. From now on in this section we assume that $\overline{F} \neq \emptyset$ (otherwise P_R is inconsistent for any α). The uncertainty with affine data perturbation occurs in many real situations. It covers many commonly used data uncertainty sets such as norm uncertainty set, box uncertainty set, and ellipsoidal uncertainty set where the set Z is a unit ball, a box, and an ellipsoid, respectively (see [14]). We must guarantee the robust feasibility, i.e., the consistency of the robust counterpart P_R, through conditions that can be expressed in terms of the data. Observe that both $\mathcal{U}$ and P_R depend on the parameter $\alpha \in \mathbb{R}_+^T$.

We first consider the uncertainty set-valued mapping $\mathcal{U}$ in (3.7) with $\alpha \in \mathbb{R}_+$ (identified here with a constant mapping) and $Z = \mathbb{B}$, where $\mathbb{B}$ denotes the closed unit ball for some norm $\|\cdot\|$ in $\mathbb{R}^{n+1}$, that is,

$$U_t := \left(\overline{a}_t, \overline{b}_t\right) + \alpha\mathbb{B}, \; t \in T. \tag{3.8}$$

Observe that if P_R is consistent for $\beta > 0$, then P_R is consistent also for γ, for any nonnegative $\gamma < \beta$. Hence, the set $\{\alpha \in \mathbb{R}_+ : P_R \text{ is consistent for } \alpha\}$ is an interval whose minimum element is 0.

The *radius of consistency* of the robust counterpart associated with $\mathcal{U}$ is

$$\rho\left(\mathcal{U}\right) := \sup\{\alpha \in \mathbb{R}_+ : P_R \text{ is consistent for } \alpha\}. \tag{3.9}$$

The supremum in (3.9) cannot be $+\infty$ since, given $t \in T$, $(0_n, 1) \in \left(\overline{a}_t, \overline{b}_t\right) + \alpha\mathbb{B}$ for a positive large enough α, in which case the corresponding problem P_R is not consistent. Moreover, this supremum may not always be attained.

The next result provides a formula for the radius of consistency which involves the so-called *hypographical set* [49] of the system $\left\{\overline{a}_t'x \geq \overline{b}_t, t \in T\right\}$, defined as

$$H\left(\overline{a},\overline{b}\right) := \text{conv}\left\{\left(\overline{a}_t,\overline{b}_t\right), t \in T\right\} + \mathbb{R}_+\left\{(0_n, -1)\right\}.$$

Since we are assuming that F_R is consistent, $\overline{F} \neq \emptyset$, so that, by the existence theorem, 0_{n+1} cannot be an interior point of the characteristic cone of $\left\{\overline{a}_t'x \geq \overline{b}_t, t \in T\right\}$, which obviously contains $H\left(\overline{a},\overline{b}\right)$. So, denoting by d the distance associated with $\|\cdot\|$, one has $d\left(0_{n+1}, \text{bd } H\left(\overline{a},\overline{b}\right)\right) = d\left(0_{n+1}, \text{cl } H\left(\overline{a},\overline{b}\right)\right)$.

Theorem 3.1.3 (Calculus of the Consistency Radius). *Let $\mathcal{U}$ be as in (3.8). Then the equation*

$$\rho\left(\mathcal{U}\right) = d\left(0_{n+1}, \text{cl } H(\overline{a},\overline{b})\right)$$

holds under any of the following conditions:

(i) $\{(\overline{a}_t, \overline{b}_t), t \in T\}$ *is a bounded set without isolated points.*
(ii) $\mathbb{B}$ *is the Euclidean unit ball and* $\{(\overline{a}_t, \overline{b}_t), t \in T\}$ *is compact.*
(iii) $\mathbb{B}$ *is a polytope and* $\{(\overline{a}_t, \overline{b}_t), t \in T\}$ *is finite.*

The inequality $\rho(\mathcal{U}) \leq d\left(0_{n+1}, \operatorname{cl} H(\overline{a}, \overline{b})\right)$ follows from the identification of the selections of $\mathcal{U}$ with perturbations of $\left(\overline{a}, \overline{b}\right)$ and the application of Theorem 6.2.2 (which is valid for any norm in $\mathbb{R}^{n+1}$). Recall that $\{(\overline{a}_t, \overline{b}_t), t \in T\}$ is compact whenever the nominal system $\left\{\overline{a}_t' x \geq \overline{b}_t, t \in T\right\}$ is continuous and that the unit ball $\mathbb{B}$ is a polytope for the norms $\|\cdot\|_\infty$ and $\|\cdot\|_1$. Example 6.2.1 below illustrates the computation of $d_2\left(0_{n+1}, \operatorname{cl} H\left(\overline{a}, \overline{b}\right)\right)$ for Example 1.1.1.

Theorem 3.1.4 (Attainability of the Consistency Radius). *If* $0^+ \overline{F}$ *is a linear subspace, then the supremum in (3.9) is attained.*

Sketch of the Proof. We assume w.l.o.g. that $\rho(\mathcal{U}) > 0$. Let $(\alpha_k)_{k=1}^\infty \subset \mathbb{R}_{++}$ be such that $\alpha_k \to \rho(\mathcal{U})$. Then, for each $k \in \mathbb{N}$ there exists $x_k \in \mathbb{R}^n$ such that

$$v_t' x_k - w_t \geq 0 \text{ for all } (v_t, w_t) \in \left(\overline{a}_t, \overline{b}_t\right) + \alpha_k \mathbb{B}, t \in T.$$

This implies that

$$\overline{a}_t' x_k - \overline{b}_t + \inf_{(c_t, d_t) \in \mathbb{B}} \alpha_k \left(c_t' x_k - d_t\right) \geq 0 \text{ for all } t \in T.$$

Let $\gamma, \delta \in \mathbb{R}_{++}$ be such that $\gamma \mathbb{B}_2 \subset \mathbb{B} \subset \delta \mathbb{B}_2$. Then

$$\inf_{(c_t, d_t) \in \mathbb{B}} \alpha_k \left(c_t' x_k - d_t\right) \geq \inf_{(c_t, d_t) \in \delta \mathbb{B}_2} \alpha_k \left(c_t' x_k - d_t\right) \geq -\delta \alpha_k \left\|(x_k, -1)\right\|_2.$$

So, one has

$$\overline{a}_t' x_k - \overline{b}_t - \delta \alpha_k \left\|(x_k, -1)\right\|_2 \geq 0 \text{ for all } t \in T. \tag{3.10}$$

Since $(x_k)_{k=1}^\infty$ is bounded, we can assume w.l.o.g. that $x_k \to \overline{x}$. Taking limits in (3.10) we get $\overline{a}_t' \overline{x} - \overline{b}_t - \delta \rho(\mathcal{U}) \left\|(\overline{x}, -1)\right\|_2 \geq 0$ for all $t \in T$, so that $\overline{x} \in F_\mathbb{R}$. Thus the supremum in (3.9) is attained (for the details, see [96, Proposition 3.7]).

Observe that, for the selection $\left(\overline{a}, \overline{b}\right)$ associated with the constraint system in Example 1.1.1, one has $0^+ \overline{F} = \{0_2\}$, so that the supremum in (3.9) is attained.

The next corollary shows that $\rho(\mathcal{U}) > 0$ for the model (3.7) whenever α is a bounded function, and its proof is a straightforward application of Theorem 3.1.3.

Corollary 3.1.1 (Consistency of the Robust Counterpart). *Let* $\mathcal{U}$ *in (3.7) be such that* $\|z\| < \mu$ *for all* $z \in Z$ *and assume that* $\left(\overline{a}, \overline{b}\right)$ *and* $\mathbb{B}$ *satisfy one of the conditions (i)–(iii) in Theorem 3.1.3. Then* P_R *is consistent for any* $\alpha \in \mathbb{R}^T_+$ *such that*

$$\sup_{t \in T} |\alpha_t| \leq \mu^{-1} d\left(0_{n+1}, \mathrm{cl}\, H\left(\overline{a}, \overline{b}\right)\right).$$

Finally in this section, we consider several issues concerning the numerical treatment of P_R. Obviously, P_R is continuous if and only if T_R is compact, in which case P_R can be solved via discretization. Concerning the relevant properties of the feasible set F_R, recalling the characterization of the properties of the feasible set of a linear system in terms of its characteristic cone and (3.3), we have that F_R is a convex body if and only if cl cone T_R is pointed and $(0_n, -1) \in$ int cone T_R, in which case P_R can be solved by means of the interior point constraint generation algorithm. Similarly, concerning the optimal set, S_R is bounded if and only if $\overline{c} \in$ int cone $\mathrm{proj}_{\mathbb{R}^n}(T_R)$ by [102, Corollary 9.3.1], in which case P_R can be solved by some central cutting plane algorithm. The next result concerns the tractability of the robust counterpart under uncertainty with affine data perturbation.

Theorem 3.1.5 (Tractability of the Robust Counterpart).   *Let* $\mathcal{U}$ *be as in (3.7) with* $\left\{\overline{a}'_t x \geq \overline{b}_t, t \in T\right\}$ *continuous,* $\alpha \in \mathcal{C}(T)$, *and* $Z \subset \mathbb{R}^{n+1}$ *compact. Then, the following statements hold:*

(i) P_R *is continuous provided it is consistent.*
(ii)   *If* Z *is a convex body and* $\alpha \in \mathbb{R}^T_{++}$, *then* P_R *satisfies the density assumption.*
(iii)   *If* $Z = \mathrm{conv}\left\{(r_i, s_i), i \in I\right\}$, *where* I *is a finite set, then*

$$F_R = \left\{x \in \mathbb{R}^n : (\overline{a}_t + \alpha_t r_i)' x \geq \overline{b}_t + \alpha_t s_i, (t, i) \in T \times I\right\}. \qquad (3.11)$$

The assumption of (ii) guarantees that P_R can be solved via grid discretization. The assumption of statement (iii) holds, e.g., whenever $Z = \mathbb{B}_\infty$, with $I = \{1, \ldots, 2^n\}$. The next example shows that (3.11) may provide a reformulation of P_R as LSIO problem with an index set of the same dimension as T.

Example 3.1.2. Consider the uncertain LSIO problem in Example 3.1.1, with an approximation error not greater than a given $\alpha \in [0, 1]$. Then, we have P_R as in (3.4). Moreover, since $\mathbb{B}_\infty = \mathrm{conv}\left\{(\pm 1, \pm 1)\right\}$, Theorem 3.1.5(iii) allows us to reformulate P_R as

$$P_R^1 : \inf_{x \in \mathbb{R}^2} -x_1 - x_2$$

$$\text{s.t.} \quad (\alpha - \cos t)\, x_1 + (\alpha - \sin t)\, x_1 \geq -1, \ t \in \left[0, \frac{\pi}{2}\right],$$

$$-(\alpha + \cos t)\, x_1 + (\alpha - \sin t)\, x_1 \geq -1, \ t \in \left[0, \frac{\pi}{2}\right],$$

Fig. 3.1 Generators of M_R

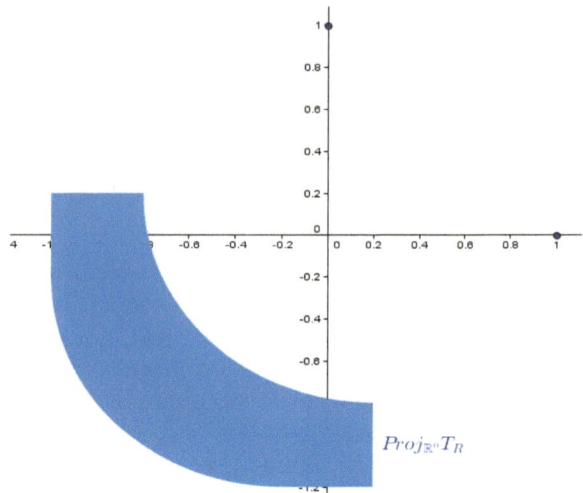

$$\left(\alpha - \cos t\right) x_1 - \left(\alpha + \sin t\right) x_1 \geq -1, \ t \in \left[0, \frac{\pi}{2}\right],$$

$$-\left(\alpha + \cos t\right) x_1 - \left(\alpha + \sin t\right) x_1 \geq -1, \ t \in \left[0, \frac{\pi}{2}\right],$$

$$x_1 \geq 0 \ (t = 2), \ x_2 \geq 0 \ (t = 3), \tag{3.12}$$

where we can replace the identical index intervals in (3.12) by $T_1 = \left[0, \frac{\pi}{2}\right]$, $T_2 = \left[2\pi, \frac{5\pi}{2}\right]$, $T_3 = \left[4\pi, \frac{9\pi}{2}\right]$, and $T_4 = \left[6\pi, \frac{13\pi}{2}\right]$, consecutively. This way we get an equivalent continuous LSIO problem P_R^2 whose feasible set F_R^2 is full dimensional and compact (as P_R^2 satisfies SCQ and $F_R^2 \subset \overline{F}$), its solution set S_R^2 is compact too, and its index set $\left(\bigcup_{i=1}^4 T_i\right) \cup \{2, 3\}$ satisfies the density assumption and is contained in the one-dimensional interval $[0, 21]$. Figures 3.1 and 3.2 represent the set of LHS vector coefficients of P_R and P_R^2 (i.e., the generators of the respective first moment cones M_R and M_R^2), for $\alpha = 0.2$ (the curves in Fig. 3.2 are labeled as the corresponding index intervals). Thus, P_R^2 could be solved via grid discretization, central cutting plane methods, and the interior point constraint generation algorithm.

Remark 3.1.1 (Antecedents and Extensions). The tractability of the robust counterparts of ordinary optimization problems has been analyzed in [14–16], etc., while robust duality theorems for LO and ordinary convex optimization have been given in [12] and [150], among others. The proofs of Theorems 3.1.1 and 3.1.2 can be found in [95, Theorem 1] and [96, Corollary 2.7]. Theorem 3.1.3(ii) is [96, Theorem 3.3], while statements (i) and (iii) can be derived from the arguments in [96, Remark 3.4]. Theorem 3.1.4 generalizes [96, Proposition 3.7] to arbitrary norms while Theorem 3.1.5 is an adaptation of [96, Proposition 4.2]. In [95, Proposition 1] it is shown that M^R is convex in the case of affinely parameterized data uncertainty

Fig. 3.2 Generators of M_R^2

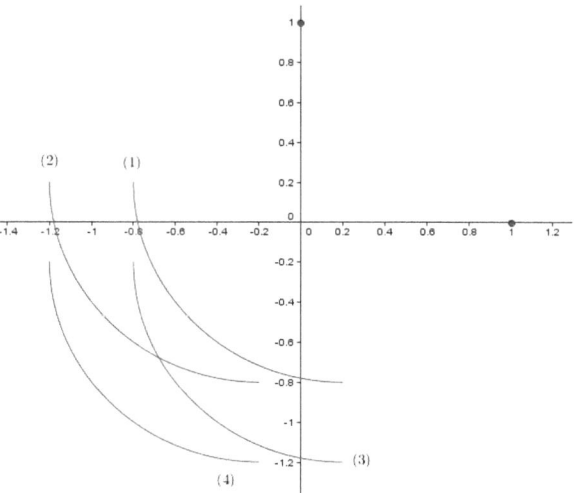

(defining this setting as in [14]) while [95, Proposition 2] shows that M^R is closed under a robust SCQ together with suitable topological requirements on the index set and the uncertainty set of the problem.

Duality theorems for robust multiobjective LSIO under uncertain constraints can be found in [96], and for robust convex optimization problems posed in locally convex spaces in [27]. The latter paper includes an application to best approximation problems with constraint data uncertainty. An extension of Theorem 3.1.5 to robust multiobjective LSIO can be found in [96, Proposition 15].

3.2 Uncertain Objective

We denote by $\overline{F}$ and $\overline{M}$ the feasible set and the first moment cone of $\left\{\overline{a}_t'x \geq \overline{b}_t,\ t \in T\right\}$ and by σ_C the support function of C. We assume that $\overline{F} \neq \emptyset$. Then the robust counterpart of P_0 is the CSIO problem

$$\inf_{x \in \mathbb{R}^n}\ \sigma_C(x) = \sup_{c \in C} c'x$$
$$\text{s.t.} \qquad \overline{a}_t'x \geq \overline{b}_t,\ t \in T,$$

or, equivalently, the LSIO problem

$$P_R : \inf_{(x,y) \in \mathbb{R}^{n+1}}\ y$$

$$\text{s.t.} \qquad -c'x + y \geq 0,\ c \in C,$$

$$\overline{a}_t'x \geq \overline{b}_t,\ t \in T, \tag{3.13}$$

so that $c_R = (0_n, 1)$, $F_R = \text{epi}\,\sigma_C \cap (\overline{F} \times \mathbb{R})$, $T_R = T \cup C$ (we can assume w.l.o.g. that $C \cap T = \emptyset$),

$$M_R = \text{cone}\{(-C) \times \{1\}) \cup \{(\overline{a}_t, 0)\,,\ t \in T\}\}$$
$$= (\overline{M} \times \{0\}) + \text{cone}\{(-C) \times \{1\}\},$$

and

$$K_R = \text{cone}\left[((-C) \times \{1\} \times \{0\}) \cup \left\{(\overline{a}_t, 0, \overline{b}_t) : t \in T\right\} \cup \{(0_n, 0, -1)\}\right].$$

Obviously, P_R is consistent if and only if $\overline{F} \cap \text{dom}\,\sigma_C \neq \emptyset$. In particular, P_R is consistent whenever C is bounded. The robust duality theory is a straightforward specification of the one exposed in Sect. 3.1, just defining the uncertainty set-valued mapping $\mathcal{U} : T \cup \{t_0\} \rightrightarrows \mathbb{R}^{n+1}$, with $t_0 \notin T$, such that $U_{t_0} = (-C) \times \{1\} \times \{0\}$ and $U_t = \left\{(\overline{a}_t, 0, \overline{b}_t)\right\}$ for all $t \in T$. Concerning the suitable numerical treatment of P_R, it depends on its relevant properties: continuity of the constraint system, density assumption, boundedness of the optimal set S_R, and boundedness and full dimensionality of F_R.

Typically, $T \subset \mathbb{R}^m$ for some $m \leq n$ and $C = \overline{c} + \alpha Z$, where $\alpha \in \mathbb{R}_{++}$ and Z is a given bounded set of $\mathbb{R}^n$. For instance, if Z is the closed unit ball for some norm $\|\cdot\|$ in $\mathbb{R}^n$, then C is a ball centered at $\overline{c}$, σ_C is finite-valued and $\sigma_C(x) = \overline{c}'x + \alpha \|x\|_*$, where $\|\cdot\|_*$ denotes the dual norm (recall that $\|\cdot\|_2$ coincides with its dual norm while $\|\cdot\|_\infty$ and $\|\cdot\|_1$ are dual to each other). Alternatively, if Z is a cartesian product of (possibly degenerate) closed intervals containing 0, C is a box not necessarily symmetric w.r.t. $\overline{c}$. In either case, we can assume w.l.o.g. that $T \subset \mathbb{R}^n$ (otherwise, we can replace T with $T \times \{0_{n-m}\}$) and $C \cap T = \emptyset$. If the nominal problem $\overline{P}$ in (2.2) is continuous, then P_R is continuous too. If, additionally, $\overline{P}$ satisfies the density assumption (e.g., T is convex), then T_R satisfies the density assumption too. Concerning the dimension of T_R, it is the maximum of the dimensions of T and C. If $x \in F$, then $(x, \max_{c \in C} c'x) \in F_R$, so that P_R is consistent (with unbounded feasible set F_R) if and only if $\overline{P}$ is consistent. Finally, concerning S_R, it is bounded if and only if

$$(0_n, 1) \in \text{int}\left[(\overline{M} \times \{0\}) + \text{cone}\{(-C) \times \{1\}\}\right]. \tag{3.14}$$

Example 3.2.1. Consider the same problem as in Example 3.1.2, but assuming now that the source of uncertainty is the objective function. Take $C = (-1, -1) + \alpha \mathbb{B}_2$, with $\alpha \in [0, 1]$. Then the robust counterpart of P_0 is

$$P_R : \inf_{(x,y) \in \mathbb{R}^3} y$$
$$\text{s.t.} \quad (1 - \alpha u_1)\,x_1 + (1 - \alpha u_2)\,x_2 + y \geq 0,\ u \in \mathbb{B}_2,$$
$$- (\cos t)\,x_1 - (\sin t)\,x_2 \geq -1,\ (t, v) \in \left[0, \tfrac{\pi}{2}\right] \times [2, 3],$$
$$x_1 \geq 0\ (t = (2, 0)),\ x_2 \geq 0\ (t = (3, 0)).$$

Obviously, P_R is continuous, the index set $\mathbb{B}_2 \cup \left[0, \frac{\pi}{2}\right] \times [2, 3] \cup \{(2, 0), (3, 0)\}$ satisfies the density assumption and is contained in the two-dimensional interval $[0, 3]^2$, and its feasible set F_R is an unbounded closed set. Moreover, since $\overline{M} \times \{0\} = \mathbb{R}^2 \times \{0\}$ and $(1, 1, 1) \in (-C) \times \{1\}$, $\left(\overline{M} \times \{0\}\right) + \text{cone} \{(-C) \times \{1\}\} = \mathbb{R}^2 \times \mathbb{R}_+$. Thus (3.14) holds and the optimal set S_R is bounded. Consequently, P_R could be solved by means of grid discretization or central cutting plane algorithms.

Remark 3.2.1 (Antecedents and Extensions). The reader is referred to the antecedents mentioned in Remark 3.1.1. In particular, [95, Corollary 2] establishes a robust duality theorem for robust LSIO with uncertain objective and constraints. Robust multiobjective LO with uncertain objective has been analyzed in [97].

Remark 3.2.2 (Two Open Problems in Robust LSIO).

1. Complexity analysis of the robust counterparts of particular types of LSIO problems.
2. Systematic analysis of the tractability of the robust counterparts of particular types of LSIO problems.

Chapter 4
Sensitivity Analysis

From now on we assume that a consistent nominal LSIO problem $\overline{P}$ as in (2.2) is given, with dual problem

$$\overline{D} : \sup_{\lambda \in \mathbb{R}^{(T)}} \sum_{t \in T} \lambda_t \overline{b}_t$$
$$\text{s.t.} \quad \sum_{t \in T} \lambda_t \overline{a}_t = \overline{c},$$
$$\lambda_t \geq 0, \ t \in T.$$

Denote by $\overline{\pi} = (\overline{c}, \overline{a}, \overline{b})$ the parameter representing $\overline{P}$ in the space of parameters $\Pi = \mathbb{R}^n \times (\mathbb{R}^n)^T \times \mathbb{R}^T$, by $M(\overline{\pi}) = \text{cone}\{\overline{a}_t, t \in T\}$ the first moment cone of $\overline{P}$, and by $K(\overline{\pi}) = \text{cone}\left\{\left(\overline{a}_t, \overline{b}_t\right), t \in T; (0_n, -1)\right\}$ the characteristic cone of $\overline{P}$. We associate with each triplet $\pi = (c, a, b) \in \Pi$ representing a perturbation of $\overline{\pi}$ the primal LSIO problem

$$P(\pi) : \inf_{x \in \mathbb{R}^n} c'x$$
$$\text{s.t.} \quad a_t'x \geq b_t, \ t \in T,$$

and its dual one

$$D(\pi) : \sup_{\lambda \in \mathbb{R}^{(T)}} \sum_{t \in T} \lambda_t b_t$$
$$\text{s.t.} \quad \sum_{t \in T} \lambda_t a_t = c,$$
$$\lambda_t \geq 0, \ t \in T.$$

Recall that in the parametric setting we denote by $\mathcal{F}(\pi)$, $\mathcal{S}(\pi)$, and $\vartheta(\pi)$ the feasible and the optimal sets of $P(\pi)$, and its optimal value, respectively. Analogously, we represent by $\mathcal{F}^D(\pi)$, $\mathcal{S}^D(\pi)$, and $\vartheta^D(\pi)$ the corresponding objects of the dual problem $D(\pi)$.

The objective of this chapter consists of determining affinity regions of the restriction of the optimal value functions ϑ (or ϑ^D) to some subspace of Π, and their estimations under three different types of perturbations of the nominal data $\overline{\pi} = (\overline{c}, \overline{a}, \overline{b})$.

M.A. Goberna and M.A. López, *Post-Optimal Analysis in Linear Semi-Infinite Optimization*, SpringerBriefs in Optimization, DOI 10.1007/978-1-4899-8044-1_4,
© Miguel A. Goberna, Marco A. López 2014

4.1   Perturbing the Objective Function

Suppose that only the objective function of $\overline{P}$ can be perturbed. Then the space of parameters is formed by the triplets of the form $\pi = \left(c, \overline{a}, \overline{b} \right)$, with parameter $c \in \mathbb{R}^n$, i.e., we can identify the subspace of parameters with $\mathbb{R}^n$.

The perturbed problems of P and D to be considered in this section are

$$P(c) : \inf_{x \in \mathbb{R}^n} c'x$$
$$\text{s.t.} \quad \overline{a}_t' x \geq \overline{b}_t, \ t \in T,$$

and

$$D(c) : \sup_{\lambda \in \mathbb{R}_+^{(T)}} \sum_{t \in T} \lambda_t \overline{b}_t$$
$$\text{s.t.} \quad \sum_{t \varepsilon T} \lambda_t \overline{a}_t = c,$$

with optimal sets $S(c)$ and $S^D(c)$, and optimal values $\vartheta(c)$ and $\vartheta^D(c)$, respectively (obviously, $P(c)$ is continuous when $\overline{P}$ is continuous). With this notation, the effective domain of the dual optimal value function ϑ^D is the first moment cone, $M(\overline{\pi})$, and the optimal values of the nominal problem $\overline{P}$ and its dual $\overline{D}$ are $\vartheta(\overline{c})$ and $\vartheta^D(\overline{c})$, respectively. We denote by $\partial \vartheta : \mathbb{R}^n \rightrightarrows \mathbb{R}^n$ the set-valued mapping which assigns to each $c \in \mathbb{R}^n$ the concave subdifferential of the optimal value function of ϑ at c.

The next result provides an estimation of the increment of $\vartheta(\overline{c}) = v\left(\overline{P}\right)$ when only the objective function of $\overline{P}$ is perturbed.

Theorem 4.1.1 (Estimation of the Optimal Value). *Let $\overline{P}$ be a consistent LSIO problem with characteristic cone $K(\overline{\pi})$. Then the following statements hold:*

(i) hypo $\vartheta^D = K(\overline{\pi})$.
(ii) hypo $\vartheta = \mathrm{cl}\, K(\overline{\pi})$.
(iii) rint $M(\overline{\pi}) \subset \mathrm{dom}\, \vartheta \subset \mathrm{cl}\, M(\overline{\pi})$.
(iv) $\partial \vartheta = S$.
(v) $c \in \mathrm{int}\, M(\overline{\pi})$ *if and only if* $S(c)$ *is bounded.*

From (i) and (ii) we conclude that the concave function ϑ is the usc hull of ϑ^D. From (iii), (iv), and [205, Theorem 23.4], we get an expression for the *directional derivative* of ϑ at a $c \in \mathrm{rint}\, M(\overline{\pi}) = \mathrm{rint\, dom}\, \vartheta$, where $\mathrm{dom}\, \vartheta = \{c \in \mathbb{R}^n : \vartheta(c) > -\infty\}$: given a direction $d \in \mathbb{R}^n$, one has

$$\vartheta'(c; d) := \lim_{\lambda \downarrow 0} \frac{\vartheta(c + \lambda d) - \vartheta(c)}{\lambda} = -\lim_{\lambda \downarrow 0} \frac{-\vartheta(c + \lambda d) + \vartheta(c)}{\lambda}$$

$$= -(-\vartheta)'(c; d) = - \sup_{u \in \partial(-\vartheta)(c)} d'u = \inf_{x \in S(c)} d'x. \quad (4.1)$$

In particular, by (v), $\vartheta'(c; d) = \min_{x \in S(c)} d'x$ whenever $c \in \text{int } M(\overline{\pi})$. Applying (4.1) to $\overline{P}$ such that $\overline{c} \in \text{rint } M(\overline{\pi})$, we conclude that a suitable estimation of $\vartheta(c)$ is given by

$$\vartheta(c) \simeq \vartheta(\overline{c}) + \inf_{x \in S(\overline{c})} \langle c - \overline{c}, x \rangle, \tag{4.2}$$

for c in the proximity of $\overline{c}$. When, $\overline{c} \in \text{int } M(\overline{\pi})$ and $\overline{P}$ has a unique optimal solution $\overline{x}$, from (iv) and (4.2) ϑ is differentiable at $\overline{c}$, with $\nabla \vartheta(c) = \{\overline{x}\}$ and $\vartheta(c) \simeq \vartheta(\overline{c}) + \langle c - \overline{c}, \overline{x} \rangle$.

The next three results extend to LSIO classical results on sensitivity analysis in LO, where exact formulas have been given for $\vartheta(c)$ in the proximity of the nominal parameter $\overline{c}$. The extension of these results to LSIO requires the introduction of suitable partition concepts.

Given a convex or concave positive homogeneous extended function f, we define the *affinity cone* of f at $z \in (\text{dom } f) \setminus \{0_n\}$, denoted by L_z, as the largest relatively open convex cone containing z on which f is *affine* (i.e., simultaneously convex and concave). An argument based on convex analysis tools shows that the collection

$$\mathcal{L}(f) := \{L_z : z \in (\text{dom } f) \setminus \{0_n\}\}$$

of affinity cones of a convex positive homogeneous function $f : \mathbb{R}^n \to \overline{\mathbb{R}}$ constitutes a partition of $(\text{dom } f) \setminus \{0_n\}$. So, we call $\mathcal{L}(f)$ the *affinity partition* of f. Since the above statement remains true for concave positive homogeneous functions and ϑ is a *superlinear* (i.e., a usc concave positive homogeneous) function, we get the following result:

Theorem 4.1.2 (Affinity Partition of the Optimal Value Function). *The class of relatively open cones $\mathcal{L}(\vartheta)$ constitutes a partition of $(\text{dom } \vartheta) \setminus \{0_n\}$ in regions where ϑ is affine.*

Thus, $L_{\overline{c}} \in \mathcal{L}(\vartheta)$ is the largest relatively open convex set (actually a convex cone) on which ϑ is affine. This information is more precise than the one given by (4.2), but it only applies on a neighborhood of $\overline{c}$ whenever $L_{\overline{c}}$ is full dimensional. This situation is approached in the following theorem.

Theorem 4.1.3 (Affinity on Neighborhoods). *ϑ is affine on some neighborhood of $\overline{c}$ if and only if $\overline{P}$ has a strongly unique solution. In such a case, if $S(\overline{c}) = \{\overline{x}\}$, then $\vartheta(c) = \langle \overline{x}, c \rangle$ for all $c \in L_{\overline{c}}$ (an open convex cone containing $\overline{c}$).*

Combining Theorem 4.1.3 with the equation

$$v(\overline{P}) = \sup_{y \in \mathbb{R}} \{y : (c, y) \in \text{cl } K(\overline{\pi})\}$$

we get the following geometric characterization of the existence of strongly unique solution.

Corollary 4.1.1. $\overline{P}$ *has a strongly unique solution if and only if the point*

$$\left(\overline{c}, \sup_{y \in \mathbb{R}} \{y : (c, y) \in \text{cl } K\,(\overline{\pi})\}\right) \in \mathbb{R}^{n+1}$$

belongs to the relative interior of a facet of cl $K\,(\overline{\pi})$.

Example 4.1.1. Consider the parametric problem

$$P\,(c) : \inf_{x \in \mathbb{R}^2} c'x$$
$$\text{s.t.} \quad -(\cos t)\,x_1 - (\sin t)\,x_2 \geq -1, \ t \in \left[0, \tfrac{\pi}{2}\right],$$
$$x_1 \geq 0\,(t = 2), \ x_2 \geq 0\,(t = 3).$$

The affinity partition $\mathfrak{L}\,(\vartheta)$ of ϑ is formed by the images by $\text{Proj}_{\mathbb{R}^2}$ of the exposed faces of hypo $\vartheta = \text{cl } K\,(\overline{\pi})$, eliminating the origin (see Fig. 4.1), so that

$$\dim L_c = \begin{cases} 1, \text{ if } c \in \left[\mathbb{R}_-^2 \cup (\mathbb{R}_+ \times \{0\}) \cup (\{0\} \times \mathbb{R}_+)\right] \setminus \{0_2\}, \\ 2, \text{ otherwise.} \end{cases}$$

The linearity cones are as follows for the three different objective functions considered in Example 1.1.1:

Case (a) $\overline{c} = (1, 1) : \vartheta \equiv 0$ on the open convex cone $L_{\overline{c}} = \mathbb{R}_{++}^2$. Since ϑ is affine on a neighborhood $L_{\overline{c}}$ of $\overline{c}$, $P\,(\overline{c})$ has a strongly unique optimal solution (the origin 0_2).

Case (b) $\overline{c} = (-1, -1) : \vartheta\,(c) = \frac{c_1 + c_2}{\sqrt{2}}$ for all $c \in L_{\overline{c}} = \mathbb{R}_{++}(-1, -1)$.

Case (c) $\overline{c} = (1, 0) : \vartheta\,(c) = 0$ for all $c \in L_{\overline{c}} = \mathbb{R}_{++}\,(1, 0)$.

Table 4.1 shows the close relationship between affinity cones of the points $c \in \mathbb{R}^2 \setminus \{0_2\}$, and the maximal optimal partitions and the optimal sets of $P\,(c)$. More precisely, all problems $P\,(c)$ corresponding to vectors c from a given affinity cone $L \in \mathfrak{L}\,(\vartheta)$ have the same maximal optimal partition. This cannot be a general rule as LSIO problems may have or not maximal optimal partition.

From the definition of $\mathfrak{L}\,(\vartheta)$, it is obvious that if $\{c^i, i \in I\} \subset L \in \mathfrak{L}\,(\vartheta)$ implies that ϑ is affine on conv $\{c^i, i \in I\}$. The next result is a counterpart of this statement involving optimal partitions (in Example 4.1.1, the problems associated with $c^i, i \in I$, have the same maximal optimal partition if and only if $\{c^i, i \in I\}$ is contained in some element of the affinity partition).

Theorem 4.1.4 (Affinity on Polytopes). *Let* $\{c^i, i \in I\} \subset \mathbb{R}^n$, *with* I *finite, be such that there exists a common optimal partition for the family of problems* $\{P\,(c^i), i \in I\}$ *(e.g., they have the same maximal optimal partition). Then* ϑ *is affine on conv* $\{c^i, i \in I\}$.

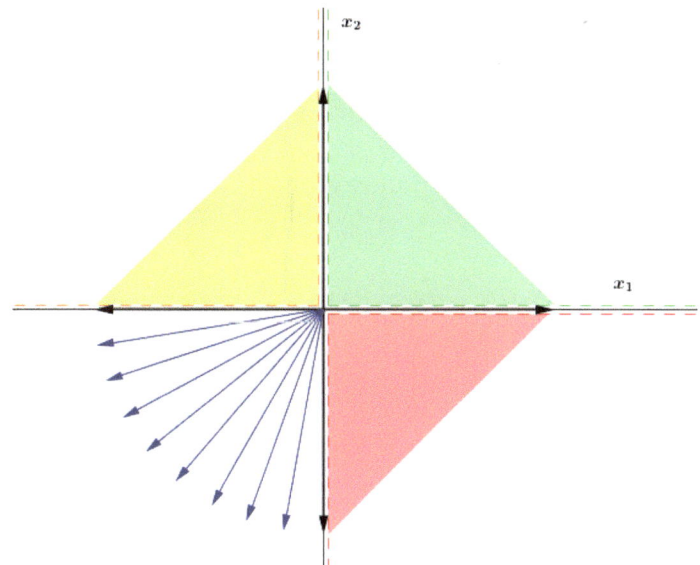

Fig. 4.1 Linearity cones partition

Table 4.1 Affinity partition in Example 4.1.1

Affinity cone	Maximal optimal partition	Optimal set
$\mathbb{R}^2_{++}$	$\left(\left[0, \frac{\pi}{2}\right], \{2, 3\}, \emptyset\right)$	$\{0_2\}$
$\{0\} \times \mathbb{R}_{++}$	$\left(\left[0, \frac{\pi}{2}\right] \cup \{2\}, \{3\}, \emptyset\right)$	$[0, 1] \times \{0\}$
$\mathbb{R}_{--} \times \mathbb{R}_{++}$	$\left(\left]0, \frac{\pi}{2}\right] \cup \{2\}, \{0, 3\}, \emptyset\right)$	$\{(1, 0)\}$
$\mathbb{R}_{--} \times \{0\}$	$\left(\left]0, \frac{\pi}{2}\right] \cup \{2\}, \{0\}, \{3\}\right)$	$\{(1, 0)\}$
$\mathbb{R}_{++}c, c \in \mathbb{R}^2_{--}$	$(T \setminus \{\alpha\}, \{\alpha\}, \emptyset), \alpha = \arctan\left(\frac{c_2}{c_1}\right)$	$\left\{\left(-\frac{c_1}{\|c\|_2}, -\frac{c_2}{\|c\|_2}\right)\right\}$
$\{0\} \times \mathbb{R}_{--}$	$\left(\left[0, \frac{\pi}{2}\right[\cup \{3\}, \left\{\frac{\pi}{2}\right\}, \{2\}\right)$	$\{(0, 1)\}$
$\mathbb{R}_{++} \times \mathbb{R}_{--}$	$\left(\left[0, \frac{\pi}{2}\right[\cup \{3\}, \left\{\frac{\pi}{2}, 2\right\}, \emptyset\right)$	$\{(0, 1)\}$
$\mathbb{R}_{++} \times \{0\}$	$\left(\left[0, \frac{\pi}{2}\right] \cup \{3\}, \{2\}, \emptyset\right)$	$\{0\} \times [0, 1]$

The duality approach to sensitivity analysis provides the next result on affinity along segments under assumptions implying the existence of optimal partition.

Theorem 4.1.5 (Affinity Along Segments). *ϑ is affine on a segment emanating from $\bar{c}$ in the direction of $d \in \mathbb{R}^n \setminus \{0_n\}$ if $\overline{P}$ and $\overline{D}$ are solvable, with $v\left(\overline{P}\right) = v\left(\overline{D}\right)$, and the problem*

$$\overline{D}_d : \sup\nolimits_{\lambda \in \mathbb{R}^{(T)}_+, \mu \in \mathbb{R}} \sum_{t \in T} \lambda_t \overline{b}_t + \mu v\left(\overline{P}\right)$$
$$s.t. \qquad \sum_{t \in T} \lambda_t \overline{a}_t + \mu \overline{c} = d,$$

is also solvable and $v\left(\overline{P}_d\right) = v\left(\overline{D}_d\right)$.

By Theorem 1.2.1, the assumptions on $\overline{D}_d$ hold whenever its corresponding primal problem

$$\overline{P}_d : \inf_{x \in \mathbb{R}^n} d'x$$
$$\text{s.t.} \quad \overline{a}'_t x \geq \overline{b}_t, \ t \in T,$$
$$\overline{c}'x = v\left(\overline{P}\right),$$

satisfies the FMCQ.

Example 4.1.2. Let $\overline{P}$ be the problem $P(-1,-1)$ of Example 4.1.1(b) and $d \in \mathbb{R}^2$ such that $0 < \|d\|_2 < 1$. Here

$$\overline{P}_d : \inf_{x \in \mathbb{R}^n} d'x$$
$$\text{s.t.} \quad -(\cos t)\, x_1 - (\sin t)\, x_2 \geq -1, \ t \in \left[0, \tfrac{\pi}{2}\right],$$
$$x_1 \geq 0, \ x_2 \geq 0, \ x_1 + x_2 = \sqrt{2}.$$

Denoting by $K(\overline{\pi})$ and K_d the characteristic cones of $\overline{P}$ and $\overline{P}_d$, we have $K_d = K(\overline{\pi}) + \text{span}\left\{\left(1, 1, \sqrt{2}\right)\right\}$, and this cone is not closed (its closure is the half-space $x_3 \leq x_1 + x_2$). So, the reader may verify that neither Theorem 4.1.3 nor Theorem 4.1.5 can be applied.

Observe that the maximal optimal partitions of $\overline{c} = (-1, -1)$ and $\overline{c} + d = (-1 + d_1, -1 + d_2)$, are $\left(T \setminus \left\{\tfrac{\pi}{4}\right\}, \left\{\tfrac{\pi}{4}\right\}, \emptyset\right)$ and $\left(T \setminus \{\alpha\}, \{\alpha\}, \emptyset\right)$, which coincide if and only if $\alpha = \arctan\left(\tfrac{-1+d_2}{-1+d_1}\right) = \tfrac{\pi}{4}$, i.e., $d \in \mathbb{R}_{++}\overline{c}$. So, Theorem 4.1.4 applies whenever $d \in \mathbb{R}_{++}\overline{c}$ to conclude that ϑ is affine on the segment $[\overline{c}, \overline{c} + d]$. Thus, ϑ is affine on the half-line $\mathbb{R}_{++}\overline{c}$ (see Fig. 4.1).

Remark 4.1.1 (Antecedents and Sources). Theorem 4.1.1 on the properties of the optimal value function has the antecedent of the duality theorem for superconsistent LSIO problems in [209], whose proof is quite schematic; a more detailed proof can be found in [102, Theorem 8.1]. Conditions for the differentiability of ϑ at $\overline{c}$ can be obtained from the identity $\vartheta(c) = -\sigma_{\overline{F}}(-c)$ and the results in [232, 233]. In particular, if $\overline{F}$ is a compact convex set with nonempty interior, then ϑ is differentiable except at the origin if and only if F is strictly convex (i.e., bd F contains no segment).

Theorems 4.1.2 and 4.1.3 have no antecedent in LO; they have been proved in [110, Proposition 2.2] and [93, Theorem 1], respectively. Finally, Theorem 4.1.5 extends a similar result of Gauvin [87] on sensitivity analysis in LO and was proved in [93, Theorem 2], while Theorem 4.1.4 was shown in [110, Proposition 4.1] and can be seen as an extension, to LSIO, of a large stream of papers on sensitivity analysis in LO from an optimal partition perspective [1,19,85,88,89,119–122,146–148,188,208]. Analogous extensions have been proposed for semidefinite optimization [117], conic optimization [230], and quadratic optimization [89].

4.2   Perturbing the RHS

Now we consider the parametric problems

$$P\left(b\right): \inf_{x \in \mathbb{R}^n} \ \overline{c}'x$$
$$\text{s.t.} \quad \overline{a}'_t x \geq b_t, \ t \in T,$$

and

$$D\left(b\right): \sup_{\lambda \in \mathbb{R}^{(T)}_+} \ \sum_{t \in T} \lambda_t b_t$$
$$\text{s.t.} \quad \sum_{t \in T} \lambda_t \overline{a}_t = \overline{c},$$

with respective optimal values $\vartheta\left(b\right)$ and $\vartheta^D\left(b\right)$. Obviously, the optimal values of the nominal problem $\overline{P}$ and its dual $\overline{D}$ are $v\left(\overline{P}\right) = \vartheta\left(\overline{b}\right)$ and $v\left(\overline{D}\right) = \vartheta^D\left(\overline{b}\right)$, respectively, as $P\left(b\right) \equiv \pi = \left(\overline{c}, \overline{a}, b\right)$ and $\overline{P} = P\left(b\right) \equiv \overline{\pi} = \left(\overline{c}, \overline{a}, \overline{b}\right)$. Notice that the first moment cone of $P\left(b\right)$ coincides with the first moment cone $M\left(\overline{\pi}\right)$ of $\overline{P}$. So, if $\overline{c} \in \text{rint } M\left(\overline{\pi}\right)$ (e.g., the feasible set $\overline{F}$ of $\overline{P}$ is bounded) and $P\left(b\right)$ is consistent, then $\vartheta^D\left(b\right) = \vartheta\left(b\right)$ by Theorem 1.2.1, i.e., ϑ^D coincides with ϑ on dom $\mathcal{F}$.

Concerning the perturbations $\overline{b} : T \to \mathbb{R}$, we consider the space of parameters, identified with $\mathbb{R}^T$, equipped with the pseudometric of the uniform convergence:

$$d_\infty\left(\psi, \varphi\right) := \sup_{t \in T} \left|\psi\left(t\right) - \varphi\left(t\right)\right|, \ \psi, \varphi \in \mathbb{R}^T.$$

The next result can be seen as a RHS counterpart of Theorem 4.1.3.

Theorem 4.2.1 (Affinity on Neighborhoods). *If ϑ is linear on a certain neighborhood of $\overline{b}$, then $\overline{D}$ has at most one optimal solution. Conversely, if there exist $x^* \in \mathcal{F}\left(\overline{b}\right), \mu > 0$ and $\gamma > 0$ such that*

(i) $\overline{c} \in A\left(x^*\right)$,
(ii) $\{\overline{a}_t, t \in T\left(x^*\right)\}$ *is a basis of* $\mathbb{R}^n$, *and*
(iii) $\overline{a}'_t x \geq \overline{b}_t + \mu$ *for all* $x \in x^* + \gamma \mathbb{B}_2$ *and* $t \notin T\left(x^*\right)$,

then ϑ is the linear function $\vartheta\left(b\right) = c'x\left(b\right)$ in a neighborhood of $\overline{b}$, where $x\left(b\right)$ is the unique solution of the system $\{\overline{a}'_t x = b_t, t \in T\left(x^*\right)\}$.

The following result guarantees the affinity of ϑ along segments under strong assumptions.

Theorem 4.2.2 (Affinity Along Segments). *ϑ is affine on a segment emanating from $\overline{b}$ in the direction of a bounded function $f \in \mathbb{R}^T \setminus \{0_T\}$ if $\overline{P}$ and $\overline{D}$ are solvable with the same optimal value, the problem*

$$P_f : \inf_{(x,y) \in \mathbb{R}^{n+1}} c'x + \vartheta \left(\overline{b} \right) y$$

$$\text{s.t.} \qquad \overline{a}'_t x + \overline{b}_t y \geq f_t, \ t \in T,$$

is also solvable and has zero duality gap with D_f, and there exists $x^* \in S\left(\overline{b}\right)$ such that either $T(x^*) = T$ or there exist two scalars μ and η such that $0 < \mu \leq \overline{a}'_t x^* - \overline{b}_t \leq \eta$ for all $t \notin T(x^*)$.

Theorem 4.2.3 (Affinity on Polytopes). *Let $\mathrm{conv}\left\{b^i, i \in I\right\}$, with I finite, be such that there exists a common optimal partition for all the problems $P\left(b^i\right), i \in I$. Then $\vartheta(b) = \vartheta^D(b)$ is affine on $\mathrm{conv}\left\{b^i, i \in I\right\}$.*

Hence, if $d \in \mathbb{R}^T$ and there exists $\varepsilon > 0$ such that $P(b + \varepsilon d)$ has the same optimal partition as P, then $\vartheta(b) = \vartheta^D(b)$ is affine on $[b, b + \varepsilon d]$.

Example 4.2.1. We revisit Example 4.1.1(b), where $T = \left[0, \frac{\pi}{2}\right] \cup \{2, 3\}$. The primal problem associated with $b \in \mathbb{R}^T$ is

$$P(b) : \inf_{x \in \mathbb{R}^2} -x_1 - x_2$$
$$\text{s.t.} \qquad -(\cos t)\, x_1 - (\sin t)\, x_2 \geq b_t, \ t \in \left[0, \frac{\pi}{2}\right],$$
$$x_1 \geq b_2, \ x_2 \geq b_3,$$

and the nominal parameter is $\overline{b} \in \mathbb{R}^T$ such that $\overline{b}_t = -1$ for all $t \in \left[0, \frac{\pi}{2}\right]$, and $\overline{b}_2 = \overline{b}_3 = 0$. Recall that $\overline{x} = \left(\frac{1}{\sqrt{2}}, \frac{1}{\sqrt{2}}\right)$ and $\left(T \setminus \left\{\frac{\pi}{4}\right\}, \left\{\frac{\pi}{4}\right\}, \emptyset\right)$ are the unique optimal solution and the maximal optimal partition of $P\left(\overline{b}\right)$, respectively.

Now we consider individual perturbations of the RHS of the constraints which preserve the optimality of $\overline{x}$.

The constraint corresponding to $t \in \left[0, \frac{\pi}{2}\right]$ is redundant in $P\left(\overline{b}\right)$, so that $\mathcal{F}\left(\overline{b} + \alpha \chi^t\right) = \mathcal{F}\left(\overline{b}\right)$ for any $\alpha \leq 0$, where χ^t denotes the characteristic function of t. Even more, $\overline{x} \in \mathcal{F}\left(\overline{b} + \alpha \chi^t\right) \subset \mathcal{F}\left(\overline{b}\right)$ for any $\alpha \leq \overline{a}'_t \overline{x}$. Thus, $S\left(\overline{b} + \alpha \chi^t\right) = \{\overline{x}\}$ for any $\alpha \leq \overline{a}'_t \overline{x}$. Repeating the argument of Example 1.2.1 one concludes that $\left(T \setminus \left\{\frac{\pi}{4}\right\}, \left\{\frac{\pi}{4}\right\}, \emptyset\right)$ is the maximal optimal partition of $P\left(\overline{b} + \alpha \chi^t\right) - D\left(\overline{b} + \alpha \chi^t\right)$ for any $\alpha \leq \overline{a}'_t \overline{x} = -\frac{\cos t + \sin t}{\sqrt{2}} = -\cos\left(t - \frac{\pi}{4}\right)$.

The constraint corresponding to $t = 2, 3$ are not redundant, but it is still true that $\left(T \setminus \left\{\frac{\pi}{4}\right\}, \left\{\frac{\pi}{4}\right\}, \emptyset\right)$ is the maximal optimal partition of $P\left(\overline{b} + \alpha \chi^t\right) - D\left(\overline{b} + \alpha \chi^t\right)$ for any $\alpha \leq \overline{a}'_t \overline{x} = -\frac{1}{\sqrt{2}}, t = 2, 3$.

Then, by Theorem 4.2.3, ϑ is affine on

$$\mathrm{conv}\left(\bigcup_{t \in T} \left\{\overline{b} + \alpha \chi^t : \alpha \leq \overline{a}'_t \overline{x}\right\}\right).$$

Remark 4.2.1 (Antecedents and Sources). Antecedents for the results in this section are the works on sensitivity analysis in LO from an optimal partition perspective mentioned in Remark 4.1.1. Theorems 4.2.1, 4.2.2, and 4.2.3 are [93, Theorem 4], [93, Theorem 5], and [110, Proposition 5.1], respectively.

4.3 Perturbing the Objective Function and the RHS

We associate with the given nominal problem $\overline{P}$ the space of parameters $\mathbb{R}^T \times \mathbb{R}^n$. The primal and dual problems associated with a perturbation $\pi = (\overline{a}, b, c)$ of the nominal parameter $\overline{\pi} = (\overline{c}, \overline{a}, \overline{b})$ are

$$P\,(c,b) : \inf_{x \in \mathbb{R}^n}\ c'x$$
$$\text{s.t.} \qquad \overline{a}'_t x \geq b_t,\ t \in T,$$

and

$$D\,(c,b) : \sup_{\lambda \in \mathbb{R}^{(T)}} \sum_{t \in T} \lambda_t b_t$$
$$\text{s.t.} \qquad \sum_{t \in T} \lambda_t \overline{a}_t = c,$$
$$\lambda_t \geq 0,\ t \in T,$$

with optimal values $\vartheta\,(c,b)$ and $\vartheta^D\,(c,b)$, respectively. In order to describe the behavior of ϑ and ϑ^D, we define a class of functions after giving a brief motivation.

Let L be a linear space and let $\varphi : L^2 \to \mathbb{R}$ be a bilinear form on L. Let $C = \text{conv}\,\{v_i, i \in I\} \subset L$ and let $q_{ij} := \varphi\,(v_i, v_j)$, $(i, j) \in I^2$. Then any $v \in C$ can be expressed as

$$v = \sum_{i \in I} \mu_i v_i,\quad \sum_{i \in I} \mu_i = 1,\ \mu \in \mathbb{R}_+^{(I)}. \tag{4.3}$$

Then we have

$$\varphi\,(v, v) = \sum_{i,j \in I} \mu_i \mu_j q_{ij}. \tag{4.4}$$

Accordingly, given $q : C \times C \to \mathbb{R}$, where $C = \text{conv}\,\{v_i, i \in I\} \subset L$, we say that q is *quadratic* on C if $\exists q_{ij} \in \mathbb{R}$, $i, j \in I$, such that (4.4) holds for all $v \in C$ satisfying (4.3).

Theorem 4.3.1 (Quadratic Behavior on Polytopes). *Let* $\{(c^i, b^i), i \in I\} \subset \mathbb{R}^n \times \mathbb{R}^T$, *with* I *finite, be such that there exists a common optimal partition for the family of problems* $P\,(c^i, b^i)$, $i \in I$. *Then* $P\,(c, b)$ *and* $D\,(c, b)$ *are solvable,* $\vartheta\,(c, b) = \vartheta^D\,(c, b)$ *on* $\text{conv}\,\{c^i : i \in I\} \times \text{conv}\,\{b^i : i \in I\}$ *and* ϑ *is quadratic on*

$\text{conv}\left\{(c^i, b^i) : i \in I\right\}$. *Moreover, if* $(c, b) \in \text{conv}\left\{c^i : i \in I\right\} \times \text{conv}\left\{b^i : i \in I\right\}$, *then* $\vartheta\left(\cdot, b\right)$ *and* $\vartheta\left(c, \cdot\right)$ *are affine on* $\text{conv}\left\{c^i : i \in I\right\}$ *and* $\text{conv}\left\{b^i : i \in I\right\}$, *respectively.*

So, given $(d, f) \in \mathbb{R}^n \times \mathbb{R}^T$, if there exists $\varepsilon > 0$ such that the problem $P\left(\left(\overline{c}, \overline{b}\right) + \varepsilon(d, f)\right)$ has the same maximal optimal partition as P, then $\vartheta(c, b) = \vartheta^D(c, b)$ is quadratic on the interval $\left[\left(\overline{c}, \overline{b}\right), \left(\overline{c}, \overline{b}\right) + \varepsilon(d, f)\right]$. Moreover, $\vartheta\left(c, \overline{b}\right)$ ($\vartheta\left(\overline{c}, b\right)$) is an affine function of c on $[\overline{c}, \overline{c} + \varepsilon d]$ (of b on $\left[\overline{b}, \overline{b} + \varepsilon f\right]$, respectively).

Example 4.3.1. Let us consider simultaneous perturbations of the objective function and RHS coefficients of the nominal problem of Examples 4.1.2 and 4.2.1. Combining the arguments there, we get that ϑ is quadratic on $\text{conv}\left\{(c^i, b^i) : i \in I\right\}$, where $\left\{c^i : i \in I\right\} = R_{++}\overline{c}$ and

$$\left\{b^i : i \in I\right\} = \text{conv}\left(\bigcup_{t \in T}\left\{\overline{b} + \alpha\chi^t : \alpha \leq \overline{a}_t'\overline{x}\right\}\right).$$

Remark 4.3.1 (Antecedents and Sources). Once again the antecedents for the results in this section are the works on sensitivity analysis in LO from an optimal partition perspective mentioned in Remark 4.1.1. Theorem 4.3.1 is [110, Proposition 6.1]. Concerning the estimations of the optimal value through directional derivatives, [234] deals with the continuous linear parametric problem $P(\lambda) = \pi(\lambda) = (\overline{c}, a(\lambda), b(\lambda))$ (i.e., $a(\lambda)$ and $b(\lambda)$ are continuous functions of t on T which depend on a parameter λ), the nominal problem being $\overline{P} = P(\overline{\lambda}) = \pi(\overline{\lambda})$. The approach of the authors extends to LSIO some results in [118]. Under the assumption of *superconsistency* of the nominal dual pair $P(\overline{\lambda}) - D(\overline{\lambda})$ (i.e., SCQ and $\overline{c} \in \text{int } M(\overline{\lambda})$), and other technical assumptions, Theorem 6 in [234] provides expressions for the right-hand and left-hand side derivatives of the optimal value function ϑ at $\overline{\lambda}$. Concerning extensions, Shapiro [210, Theorem 3.2] provides sufficient conditions for the right-hand side differentiability of the optimal value function for differentiable CSIO problems and gives an explicit formula [210, (3.7)] for this derivative.

Remark 4.3.2 (Some Open Problems in Sensitivity Analysis of LSIO Problems).

1. Sensitivity analysis of LSIO problems under perturbations of a.
2. Sensitivity analysis of LSIO problems under perturbations of the triple (c, a, b).
3. Numerical methods for the computation of affinity regions under perturbations of c, b, and the couple (c, b).

Chapter 5
Qualitative Stability Analysis

5.1 Irrestricted Stability

In this section we analyze the continuity properties of $\mathcal{F}, \mathcal{S} : \Pi \rightrightarrows \mathbb{R}^n$ and $\vartheta : \Pi \to \overline{\mathbb{R}}$ at a given nominal problem $\overline{\pi} = (\overline{c}, \overline{a}, \overline{b})$ under arbitrary perturbations of all the data. The first contributions to the stability of continuous LSIO and general LSIO problems were published in the 1980s (e.g., [29, 82]) and 1990s [103, 104], respectively, and the main results in these works are gathered in the monograph [102].

We have defined in Sect. 2.1.5 the three main concepts involved in the qualitative stability analysis of a set-valued mapping $\mathcal{M} : Y \rightrightarrows X$ at $\overline{y} \in \text{dom} \, \mathcal{M}$ (equivalently, $\mathcal{M}(\overline{y}) \neq \emptyset$): lower and upper semicontinuity, and closedness. We introduce here some related concepts.

- Again when Y and X are pseudometric spaces, we can define the *inner limit*

$$\liminf_{y \to \overline{y}} \mathcal{M}(y) := \left\{ \begin{array}{c} x \in X : \forall (y_r)_{r=1}^\infty \to \overline{y} \text{ an associated } r_0 \text{ exists} \\ \text{such that } (y_r)_{r=r_0}^\infty \subset \text{dom} \, \mathcal{M}, \\ \text{and } \exists x_r \in \mathcal{M}(y_r) \; \forall r \geq r_0 \text{ such that } x_r \to x \end{array} \right\},$$

and the *outer limit*

$$\limsup_{y \to \overline{y}} \mathcal{M}(y) := \left\{ \begin{array}{c} x \in X : \exists (y_r)_{r=1}^\infty \to \overline{y} \text{ and } (x_r)_{r=1}^\infty, \; x_r \in \mathcal{M}(y_r), \\ \text{such that } x_r \to x \end{array} \right\}.$$

Following [207], we say that $\mathcal{M}$ is *outer semicontinuous* (osc) at $\overline{y}$ when $\mathcal{M}(\overline{y}) \supset \limsup_{y \to \overline{y}} \mathcal{M}(y)$. Similarly, $\mathcal{M}$ is *inner semicontinuous* (isc) at $\overline{y}$ if $\mathcal{M}(\overline{y}) \subset \liminf_{y \to \overline{y}} \mathcal{M}(y)$. Finally, we say that $\mathcal{M}$ is *continuous* at $\overline{y} \in \text{dom} \, \mathcal{M}$ if $\mathcal{M}$ is isc and osc at $\overline{y}$, i.e.,

$$\liminf_{y \to \overline{y}} \mathcal{M}(y) = \limsup_{y \to \overline{y}} \mathcal{M}(y) = \mathcal{M}(\overline{y}),$$

M.A. Goberna and M.A. López, *Post-Optimal Analysis in Linear Semi-Infinite Optimization*, SpringerBriefs in Optimization, DOI 10.1007/978-1-4899-8044-1_5, © Miguel A. Goberna, Marco A. López 2014

in which case we write

$$\lim_{y \to \overline{y}} \mathcal{M}(y) = \mathcal{M}(\overline{y}).$$

Inner semicontinuity of $\mathcal{M}$ at $\overline{y}$ is equivalent to the lower semicontinuity of $\mathcal{M}$ at $\overline{y}$ (in the sense of Berge), whereas the outer semicontinuity of $\mathcal{M}$ at $\overline{y}$ is equivalent to the closedness of $\mathcal{M}$ at $\overline{y}$.

- We also say that $\mathcal{M}$ is *inner semicontinuous at* $(\overline{y}, \overline{x}) \in$ gph $\mathcal{M}$ if for every open set $U \ni \overline{x}$ there exists $V \in \mathfrak{N}_{\overline{y}}$ such that $\mathcal{M}(y) \cap U \neq \emptyset$ for all $y \in V$, that is, $\overline{x} \in \lim\inf_{y \to \overline{y}} \mathcal{M}(y)$. Clearly, $\mathcal{M}$ is inner semicontinuous at $\overline{y}$ if and only if $\mathcal{M}$ is inner semicontinuous at $(\overline{y}, x)$ for every $x \in \mathcal{M}(\overline{y})$.

We have already mentioned that $\mathcal{F}$ is closed at any $\pi \in$ dom $\mathcal{F}$, i.e., gph $\mathcal{F}$ is closed. The fundamental result relative to $\mathcal{F}$ characterizes the lower semicontinuity of $\mathcal{F}$ at $\overline{\pi}$ in a variety of ways, which involve the previous concepts as well as the following one:

- $\mathcal{F}$ is *stable in Tuy's sense* at $\overline{\pi}$ if $0_T \notin$ bd $\{G(\mathbb{R}^n) - \mathbb{R}_+^T\}$, where $G(x)(.) := s(x,.)$, $x \in \mathbb{R}^n$ (here $s(x,.)$ is the slack function at x, and $\mathbb{R}^T$ is assumed to be equipped with the topology of the uniform convergence on T).

Theorem 5.1.1 (Lower Semicontinuity of the Feasible Set). *Let $\overline{\pi} \in$ dom $\mathcal{F}$. The following statements are equivalent*:

   (i) $\mathcal{F}$ *is lsc at* $\overline{\pi}$;
  (ii) $\overline{\pi} \in$ int dom $\mathcal{F}$;
 (iii) *sufficiently small perturbations of the RHS of* $\overline{\pi}$ *preserve its feasibility*;
  (iv) $\overline{\pi}$ *satisfies the SSCQ*;
   (v) $0_{n+1} \notin$ cl $C(\overline{\pi})$, *where* $C(\overline{\pi}) =$ conv $\left\{ (\overline{a}_t, \overline{b}_t), t \in T \right\}$;
  (vi) *there exists* $V \in \mathfrak{N}_{\overline{\pi}}$ *such that* dim $\mathcal{F}(\pi) =$ dim $\mathcal{F}(\overline{\pi})$ *for all* $\pi \in V$ (*dimensional stability*);
 (vii) *there exists* $V \in \mathfrak{N}_{\overline{\pi}}$ *such that* aff $\mathcal{F}(\pi) =$ aff $\mathcal{F}(\overline{\pi})$ *for all* $\pi \in V$ (*affine hull immobility*);
(viii) $\overline{\pi}$ *is stable in Tuy's sense at* $\overline{\pi}$;
  (ix) $\mathcal{F}$ *is continuous at* $\overline{\pi}$.
       Moreover, in the case that $0_n \notin$ bd conv $\{\overline{a}_t, t \in T\}$, *the following condition is added to the list*:
   (x) *there exists* $V \in \mathfrak{N}_{\overline{\pi}}$ *such that* $\mathcal{F}(\pi)$ *is homeomorphic to* $\mathcal{F}(\overline{\pi})$ *for all* $\pi \in V$ (*topological stability*).

In [207] it is said that "upper semicontinuity differs from our outer semicontinuity and is seriously troublesome in its narrowness." This is why the characterization of the upper semicontinuity of $\mathcal{F}$ at $\overline{\pi}$ is a hard problem, conceptually solved in [45] developing some ideas in [104].

We have already stated that upper semicontinuity hardly holds when the image is non-compact. In fact it requires that the perturbed feasible sets differ from the nominal one in a uniformly bounded manner, as Theorem 5.1.2 below shows. A preliminary step is the following lemma:

Lemma 5.1.1 (Uniform Boundedness of the Feasible Set).   *Let $\overline{\pi} \in$ dom $\mathcal{F}$. If $\mathcal{F}(\overline{\pi})$ is bounded, $\mathcal{F}$ is uniformly bounded in some neighborhood of $\overline{\pi}$.*

Sketch of the Proof.   Otherwise, there will exist a sequence $(\pi_r)_{r=1}^{\infty}$ converging to $\overline{\pi}$ and $x_r \in \mathcal{F}(\pi_r)$ such that $\|x_r\|_2 \geq r,\ r = 1, 2, \ldots$. Then, if d is an accumulation point of $(x_r / \|x_r\|_2)_{r=1}^{\infty}$, it can easily be proved that $d \in 0^+\mathcal{F}(\overline{\pi})$ and this contradicts the assumed boundedness of $\mathcal{F}(\overline{\pi})$.

The following theorem is Theorem 3.1 in [104].

Theorem 5.1.2 (Upper Semicontinuity of the Feasible Set). $\mathcal{F}$ *is usc at $\overline{\pi} \in$* dom $\mathcal{F}$ *if and only if there exist $\rho > 0$ and a neighborhood $V \in \mathfrak{N}_{\overline{\pi}}$ such that*

$$\mathcal{F}(\pi)\setminus\rho\mathbb{B}_2 \subset \mathcal{F}(\overline{\pi})\setminus\rho\mathbb{B}_2 \text{ for all } \pi \in V. \tag{5.1}$$

Example 5.1.1.   Let us consider the parameter $\overline{\pi}$, in $\mathbb{R}^2$, whose constraint system is

$$\overline{\sigma} = \{tx_1 + x_2 \geq -|t|,\ t \in \mathbb{R}\}.$$

Obviously, $\mathcal{F}(\overline{\pi}) = [-1, 1] \times \mathbb{R}_+$ and

$$\operatorname{cl} K(\overline{\pi}) = \operatorname{cone}\{(-1, 0, -1), (1, 0, -1), (0, 1, 0)\}. \tag{5.2}$$

If $d_{\infty}(\overline{\pi}, \pi) < +\infty$ one can easily prove that $x_1 \in [-1, 1]$ for all $x \in \mathcal{F}(\pi)$, and if $d_{\infty}(\overline{\pi}, \pi) < 1$ and

$$\rho := \frac{2d_{\infty}(\overline{\pi}, \pi)}{1 - d_{\infty}(\overline{\pi}, \pi)},$$

then $x_2 \geq -\rho$ for all $x \in \mathcal{F}(\pi)$ (take $t = 0$). Therefore, $\mathcal{F}(\pi)\setminus(\rho + 1)\mathbb{B}_2 \subset \mathcal{F}(\overline{\pi})\setminus(\rho + 1)\mathbb{B}_2$ for all π such that $d_{\infty}(\overline{\pi}, \pi) < 1$, and Theorem 5.1.2 applies to conclude that $\mathcal{F}$ is usc at $\overline{\pi}$.

Unfortunately, the characterization of the upper semicontinuity of $\mathcal{F}$ at $\overline{\pi}$ given in Theorem 5.1.2 does not rely on the coefficients of the nominal problem $\overline{\pi}$. As positive counterpart, it has a straightforward consequence:

Corollary 5.1.1.   *Let $\overline{\pi} \in$ dom $\mathcal{F}$. If $\mathcal{F}(\overline{\pi})$ is bounded, then $\mathcal{F}$ is usc at $\overline{\pi}$.*

To go ahead, let us observe that the upper semicontinuity of $\mathcal{F}$ at $\overline{\pi}$ strongly depends on the coefficients ($\overline{a}$ and $\overline{b}$) as it becomes obvious from Example 3.6 in [104]. This example considers the equivalent representation of the constraint system of $\overline{\pi}$

$$\overline{\sigma} := \{k\overline{a}'_t x \geq k\overline{b}_t,\ (t, k) \in T \times \mathbb{N}\},$$

whose index set is of the same cardinality than T (when T is infinite), and it shows that $\mathcal{F}$ is trivially usc at $\overline{\sigma}$. So, by enlarging the coefficients we *forced* $\mathcal{F}$ to be usc. This observation motivates the introduction in [45] of the *reinforced system* associated with $\overline{\pi}$ which is given by

$$\overline{\sigma}^{re} := \{a'x \geq b,\ (a,b) \in 0^+ \operatorname{cl} C(\overline{\pi})\},$$

and that constitutes the main tool in the analysis of the upper semicontinuity of $\mathcal{F}$.

Taking into account the sequential interpretation of the recession directions, and thanks to the Farkas Lemma, we have $0^+ \operatorname{cl} C(\overline{\pi}) \subset \operatorname{cl} K(\overline{\pi})$, and the set of solutions of $\overline{\sigma}^{re}$, represented by $\mathcal{F}^{re}(\overline{\pi})$, contains $\mathcal{F}(\overline{\pi})$. We denote by $K^{re}(\overline{\pi})$ the characteristic cone of $\overline{\sigma}^{re}$. It is easy to verify [45, Lemma 3.4] that if $d_\infty(\overline{\pi}, \pi) < \infty$ we have $\overline{\sigma}^{re} = \sigma^{re}$ as $0^+ \operatorname{cl} C(\overline{\pi}) = 0^+ \operatorname{cl} C(\pi)$, and that $K^{re}(\overline{\pi})$ is closed [45, Corollary 5.2].

By using the reinforced system we can establish the following sufficient condition [45, Corollary 3.5] for the upper semicontinuity of $\mathcal{F}$ when $\mathcal{F}(\overline{\pi})$ is not bounded:

Corollary 5.1.2. *Let $\overline{\pi} \in \operatorname{dom} \mathcal{F}$. If $\mathcal{F}^{re}(\overline{\pi}) \backslash \mathcal{F}(\overline{\pi})$ is bounded, then $\mathcal{F}$ is usc at $\overline{\pi}$.*

Sketch of the Proof. If $\rho > 0$ is such that $\mathcal{F}^{re}(\overline{\pi}) \backslash \mathcal{F}(\overline{\pi}) \subset \rho \mathbb{B}_2$ and $d_\infty(\overline{\pi}, \pi) < \infty$, one has

$$\mathcal{F}(\pi) \backslash \mathcal{F}(\overline{\pi}) \subset \mathcal{F}^{re}(\pi) \backslash \mathcal{F}(\overline{\pi}) = \mathcal{F}^{re}(\overline{\pi}) \backslash \mathcal{F}(\overline{\pi}) \subset \rho \mathbb{B}_2,$$

and $\mathcal{F}(\pi) \backslash \rho \mathbb{B}_2 \subset \mathcal{F}(\overline{\pi}) \backslash \rho \mathbb{B}_2$ for all π to a finite distance from $\overline{\pi}$. Now, Theorem 5.1.2 applies.

The sufficient condition established in this corollary is not necessary as the same Example 5.1.1 shows. Remember that for the problem $\overline{\pi}$ studied there, $\mathcal{F}(\overline{\pi}) = [-1, 1] \times \mathbb{R}_+$ and $\mathcal{F}$ was usc at $\overline{\pi}$. Moreover, straightforward considerations lead us to

$$\operatorname{cl} C(\overline{\pi}) = C(\overline{\pi}) = \{y \in \mathbb{R}^3 : y_2 = 1 \text{ and } y_3 \leq -|y_1|\},$$

and so,

$$0^+ \operatorname{cl} C(\overline{\pi}) = \{z \in \mathbb{R}^3 : z_2 = 0 \text{ and } z_3 \leq -|z_1|\}$$
$$= \operatorname{cone}\{(-1, 0, -1), (1, 0, -1)\}.$$

Therefore,

$$\overline{\sigma}^{re} = \{-x_1 \geq -1,\ x_1 \geq -1\} \text{ and } \mathcal{F}^{re}(\overline{\pi}) = [-1, 1] \times \mathbb{R},$$

and it turns out that $\mathcal{F}^{re}(\overline{\pi}) \backslash \mathcal{F}(\overline{\pi})$ is unbounded. Moreover

$$K^{re}(\overline{\pi}) = \operatorname{cone}\{(-1, 0, -1), (1, 0, -1)\}. \tag{5.3}$$

Indeed, according to [45, Theorem 4.5], the last example illustrates the only case when the sufficient condition of Corollary 5.1.2 fails to be necessary: $0^+ \mathcal{F}^{re}(\overline{\pi}) = \mathbb{R}\{u\}$ and $0^+ \mathcal{F}(\overline{\pi}) = \mathbb{R}_+\{u\}$ for some $u \neq 0_n$. The following theorem [45, Theorem 5.3] covers all the possibilities, and it is also based on the use of the reinforced system.

Theorem 5.1.3 (Upper Semicontinuity of $\mathcal{F}$ and the Reinforced System). *If $\mathcal{F}(\overline{\pi})$ is unbounded, two cases are possible:*

(i) If $\mathcal{F}(\overline{\pi})$ contains at least one line (i.e., if $\dim\{\overline{a}_t,\ t \in T\} < n$), then $\mathcal{F}$ is usc at $\overline{\pi}$ if and only if $K^{re}(\overline{\pi}) = \text{cl } K(\overline{\pi})$.

(ii) Otherwise, if w is the sum of a certain basis of $\mathbb{R}^n$ contained in $\{\overline{a}_t, t \in T\}$, then $\mathcal{F}$ is usc at $\overline{\pi}$ if and only if there exists $\beta \in \mathbb{R}$ such that

$$\text{cone}\left(K^{re}(\overline{\pi}) \cup \{(w, \beta)\}\right) = \text{cone}\left(\text{cl } K(\overline{\pi}) \cup \{(w, \beta)\}\right). \tag{5.4}$$

Let us apply Theorem 5.1.3 to Example 5.1.1. Since $\mathcal{F}(\overline{\pi})$ contains no line, we are in case (ii), and we shall take the basis $\{\overline{a}_{-1}, \overline{a}_1\} \subset \{\overline{a}_t, t \in T\}$. Obviously $w = \overline{a}_{-1} + \overline{a}_1 = (0, 2)$ and, according to (5.2) and (5.3), (5.4) holds for $\beta = 0$. This allows us to conclude the upper semicontinuity of $\mathcal{F}$ at $\overline{\pi}$.

The following mappings are closely related to $\mathcal{F}$:

- The *boundary mapping* $\mathcal{B} : \Pi \rightrightarrows \mathbb{R}^n$ associating with each $\pi \in \Pi$ the set $\mathcal{B}(\pi) := \text{bd } \mathcal{F}(\pi)$.
- The *extreme points set mapping* $\mathcal{E} : \Pi \rightrightarrows \mathbb{R}^n$ associating with each $\pi \in \Pi$ the set $\mathcal{E}(\pi)$ of extreme points of $\mathcal{F}(\pi)$.

Let $\overline{\pi} \in \text{dom } \mathcal{F}$ be such that $\mathcal{F}(\overline{\pi}) \neq \mathbb{R}^n$. The following diagram summarizes the stability properties of $\mathcal{B}$ and the existing relationships with the corresponding properties of $\mathcal{F}$:

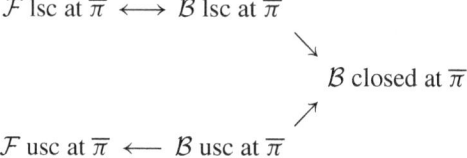

The equivalence $\mathcal{F}$ lsc at $\overline{\pi}$ $\Leftrightarrow$ $\mathcal{B}$ lsc at $\overline{\pi}$ means that the latter condition could be aggregated to the characterizations of the lower semicontinuity of $\mathcal{F}$ in Theorem 5.1.1, under the mild assumption that $\mathcal{F}(\overline{\pi}) \neq \mathbb{R}^n$. Moreover, $\mathcal{B}$ lsc at $\overline{\pi}$ $\Leftrightarrow$ $\mathcal{B}$ closed at $\overline{\pi}$ if $\dim \mathcal{F}(\overline{\pi}) = n$ and $\mathcal{B}$ usc at $\overline{\pi}$ $\Leftrightarrow$ $\mathcal{B}$ closed at $\overline{\pi}$ if $\mathcal{F}(\overline{\pi})$ is bounded.

The extreme points set mapping $\mathcal{E}$ is remarkably unstable unless the following CQ holds: $\overline{\pi}$ is *non-degenerate* if $|T(x)| < n$ for all $x \in \mathcal{B}(\overline{\pi}) \setminus \mathcal{E}(\overline{\pi}) = (\text{bd } \mathcal{F}(\overline{\pi})) \setminus (\text{extr } \mathcal{F}(\overline{\pi}))$. In the next diagram $\overline{\pi}_H := (\overline{c}, \overline{a}, 0)$ consists of replacing $\overline{b}$ in $\overline{\pi}$ by the null function, so that the constraint system of $\overline{\pi}_H$ is the homogeneous system $\overline{\sigma}_H := \{\overline{a}'_t x \geq 0, \ t \in T\}$. If $|T| \geq n$, $\mathcal{E}(\overline{\pi}) \neq \emptyset$, and $|\mathcal{F}(\overline{\pi})| > 1$ (the most difficult case), then one has:

$$\mathcal{F} \text{ lsc at } \overline{\pi} \longleftrightarrow \mathcal{E} \text{ lsc at } \overline{\pi}$$
$$\downarrow^{(1)}$$
$$(4)$$
$$\mathcal{E} \text{ closed at } \overline{\pi} \longrightarrow \overline{\pi} \text{ non-deg.}$$
$$^{(2)} \downarrow\uparrow \ ^{(3)}$$
$$(5)$$
$$\mathcal{E} \text{ usc at } \overline{\pi} \longrightarrow \overline{\pi} \ \& \ \overline{\pi}_H \text{ non-deg.}$$

The equivalence $\mathcal{F}$ lsc at $\overline{\pi}$ $\Leftrightarrow$ $\mathcal{E}$ lsc at $\overline{\pi}$ gives another condition to be added to the list in Theorem 5.1.1 under the mild assumptions that $|T| \geq n$ (superfluous in LSIO) and $\mathcal{F}(\overline{\pi})$ contains more than one point but not complete lines. The implications in the above diagram, with the exception of (5), hold under some additional assumptions: (1) $\mathcal{F}(\overline{\pi})$ is strictly convex; (2) $\mathcal{F}(\overline{\pi})$ is bounded; (3) $\{\overline{a}_t, t \in T\}$ is bounded; and (4) $\mathcal{F}$ is lsc at $\overline{\pi}$. The converse statements of (4) and (5) are true if $|T| < \infty$.

The known continuity properties of the optimal set mapping $\mathcal{S}$ and the optimal value function ϑ are gathered in the next two theorems. Through these results the reader can appreciate how strong the influence of the lower semicontinuity of $\mathcal{F}$ is.

Theorem 5.1.4 (Stability of the Optimal Set). *Let $\overline{\pi} \in$ dom $\mathcal{S}$. Then, the following statements hold:*

(i) *$\mathcal{S}$ is closed at $\overline{\pi}$ if and only if either $\mathcal{F}$ is lsc at $\overline{\pi}$ or $\mathcal{F}(\overline{\pi}) = \mathcal{S}(\overline{\pi})$.*
(ii) *$\mathcal{S}$ is lsc at $\overline{\pi}$ if and only if $\mathcal{F}$ is lsc at $\overline{\pi}$ and $\mathcal{S}(\overline{\pi})$ is a singleton set.*
(iii) *If $\mathcal{S}$ is usc at $\overline{\pi}$, then $\mathcal{S}$ is closed at $\overline{\pi}$. The converse is true whenever $\mathcal{S}(\overline{\pi})$ is bounded.*

Theorem 5.1.5 (Stability of the Optimal Value). *Let $\overline{\pi} \in$ dom $\mathcal{F}$. Then, the following statements hold:*

(i) *If $\mathcal{S}(\overline{\pi})$ is a nonempty compact set, then ϑ is lsc at $\overline{\pi}$. The converse statement holds if $\vartheta(\overline{\pi}) \neq -\infty$.*
(ii) *ϑ is usc at $\overline{\pi}$ if and only if $\mathcal{F}$ is lsc at $\overline{\pi}$.*
(iii) *If $\mathcal{S}(\overline{\pi})$ is a nonempty compact set and $\mathcal{F}$ is lsc at $\overline{\pi}$, then ϑ is Lipschitz continuous at $\overline{\pi}$, i.e., there exist $V \in \mathfrak{N}_{\overline{\pi}}$ and $L > 0$ such that*

$$|\vartheta(\pi_1) - \vartheta(\pi_2)| \leq L d(\pi_1, \pi_2) \text{ for all } \pi_1, \pi_2 \in V.$$

The latter two results involve the following conditions aside the lower semicontinuity of $\mathcal{F}$ at $\overline{\pi}$ discussed above: (a) $\mathcal{S}(\overline{\pi}) = \mathcal{F}(\overline{\pi})$, (b) $\mathcal{S}(\overline{\pi})$ is a nonempty compact set, and (c) $\mathcal{S}(\overline{\pi})$ is a singleton set. These conditions admit geometric interpretations in terms of the data although they cannot easily be checked in practice:

(a) By the non-homogeneous Farkas lemma, $\mathcal{S}(\overline{\pi}) = \mathcal{F}(\overline{\pi})$ if and only if $(\overline{c}, \vartheta(\overline{\pi}))$ belongs to the lineality of cl $K(\overline{\pi})$.

(b) $\mathcal{S}(\overline{\pi})$ is a nonempty compact set if and only if $\overline{c} \in \text{int } M(\overline{\pi})$. A sufficient condition is the existence of some feasible solution λ of the dual problem such that span $\{\overline{a}_t, \ t \in \sigma(\lambda)\} = \mathbb{R}^n$.

(c) If $\overline{x} \in \mathcal{F}(\overline{\pi})$ satisfies $\overline{c} \in \text{int } A(\overline{x})$, then $\overline{x}$ is a strongly unique solution of $\overline{\pi}$.

Example 5.1.2. Denote by π_1, π_2 and π_3 the corresponding parameters in Example 1.1.1, with cost vectors $c^1 = (1, 1)$, $c^2 = (-1, -1)$, and $c^3 = (1, 0)$, respectively. Let $\overline{\pi}$ be any of these three parameters. Since the SSCQ holds and $0_2 \in \text{int conv}\{\overline{a}_t, \ t \in T\}$, $\mathcal{F}$ satisfies all the properties listed in Theorem 5.1.1, in particular it is lsc at $\overline{\pi}$. Moreover, since the nominal feasible set is compact, $\mathcal{F}$ is also usc at $\overline{\pi}$. Consequently, the boundary mapping $\mathcal{B}$ is lsc and closed at $\overline{\pi}$ (the upper semicontinuity also holds, but it is not consequence of the general theory). Concerning the extreme point set mapping $\mathcal{E}$, it is lsc at $\overline{\pi}$ but $\mathcal{F}(\overline{\pi})$ is not strictly convex, so that the upper semicontinuity and closedness of $\mathcal{E}$ at $\overline{\pi}$ must be justified in terms of the data.

Now we apply Theorem 5.1.4. Since $\mathcal{F}$ is lsc at $\overline{\pi}$, the optimal set mapping $\mathcal{S}$ is closed at $\overline{\pi}$. This, together with the boundedness of $\mathcal{S}(\overline{\pi})$, guarantees that $\mathcal{S}$ is usc at $\overline{\pi}$. The primal problems of π_1 and π_2 have unique optimal solution, so that $\mathcal{S}$ is lsc at that parameters while the primal problem of π_3 has multiple optimal solutions and so $\mathcal{S}$ fails to be lsc at π_3.

Finally, concerning the optimal value mapping ϑ, by Theorem 5.1.5, it is lsc and usc at $\overline{\pi}$ as $\mathcal{S}(\overline{\pi})$ is a nonempty compact set and $\mathcal{F}$ is lsc at $\overline{\pi}$.

Example 5.1.3 (Uniform Approximation). The best uniform approximation of an uncertain function $\varphi : T \to \mathbb{R}$, where T is a compact interval in $\mathbb{R}$, by polynomials of degree less than n can be formulated as an uncertain unconstrained minmax problem as follows:

$$P_0 : \inf_{x \in \mathbb{R}^n} \left\{ \sup_{t \in T} \left| \varphi(t) - \sum_{i=1}^{n} t^{i-1} x_i \right| \right\}.$$

Introducing a new variable $x_{n+1} := \sup_{t \in T} \left| \varphi(t) - \sum_{i=1}^{n} t^{i-1} x_i \right|$, P_0 can be reformulated as an uncertain LSIO problem as follows:

$$P_0 : \inf_{(x, x_{n+1}) \in \mathbb{R}^{n+1}} x_{n+1}$$
$$\text{s.t.} \qquad \left| \varphi(t) - \sum_{i=1}^{n} t^{i-1} x_i \right| \leq x_{n+1}, \ t \in T.$$

We built up a parametric model for P_0 by assuming the existence of a nominal function $\overline{\varphi} : T \to \mathbb{R}$ to be approximated. The corresponding nominal approximation problem can be formulated as

$$P_{\overline{\varphi}} : \inf_{(x, x_{n+1}) \in \mathbb{R}^{n+1}} x_{n+1}$$
$$\text{s.t.} \qquad \sum_{i=1}^{n} t^{i-1} x_i + x_{n+1} \geq \overline{\varphi}(t), \ t \in T,$$
$$-\sum_{i=1}^{n} t^{i-1} x_i + x_{n+1} \geq -\overline{\varphi}(t), \ t \in T.$$

We denote by Π the space of arbitrary perturbations of all the data in $P_{\overline{\varphi}}$. Let $\mathcal{F}$, $\mathcal{S}$, and ϑ be the corresponding feasible set, optimal set, and optimal value mappings associated with $P_{\overline{\varphi}}$. We are exclusively interested in those perturbations of the data in $P_{\overline{\varphi}}$ preserving the coefficient of the variables and such that the RHS functions in both blocks of constraints, the result of perturbing $\overline{\varphi}$ and $-\overline{\varphi}$, have zero-sum. In other words, we deal with perturbations consisting of replacing $\overline{\varphi} : T \to \mathbb{R}$ in $P_{\overline{\varphi}}$ with another function $\varphi : T \to \mathbb{R}$ of the same type. So, possible spaces of perturbations are the space $\Pi_1 := \mathcal{C}(T)$ of real-valued continuous functions on T when $\overline{\varphi}$ is continuous, the space $\Pi_2 := \ell_\infty(T)$ of bounded functions on T when $\overline{\varphi}$ is bounded, and the space $\Pi_3 := \mathbb{R}^T$ when $\overline{\varphi}$ is unbounded, with $\Pi_1 \subset \Pi_2 \subset \Pi_3$. All these spaces (that can be seen as subspaces of Π in an obvious way) are equipped with the supremum metric d_∞ describing the uniform convergence on T. If $\varphi \in \Pi_1$, then P_φ has a unique optimal solution, i.e., $\mathcal{S}(\varphi)$ is a singleton set (see, e.g., [197, Theorem 7.6] or [70]). If $\overline{\varphi}$ is unbounded and $\varphi \in \Pi_3$ satisfies $d_\infty(\varphi, \overline{\varphi}) < \infty$, then φ is unbounded too and $d_\infty(\varphi, p) = +\infty$ for any polynomial p, so that P_φ is inconsistent. This means that $\mathcal{F} \equiv \emptyset, \mathcal{S} \equiv \emptyset$, and $\vartheta \equiv +\infty$ on $\{\varphi \in \Pi_3 : d_\infty(\varphi, \overline{\varphi}) < \infty\} \in \mathfrak{N}_{\overline{\varphi}}$. Thus we have just to consider the continuity properties of the restrictions of $\mathcal{F}$, $\mathcal{S}$, and ϑ to Π_1 and Π_2.

Let $\overline{\varphi} \in \Pi_i$, $i = 1, 2$. The requirement that the admissible perturbations of the RHS function of $P_{\overline{\varphi}}$ preserve the zero-sum condition obliges to apply carefully the previous results, which only provide sufficient conditions in the present framework: if one of the relevant mappings, $\mathcal{F}$, $\mathcal{S}$, and ϑ, satisfies a certain stability property at $\overline{\varphi}$ for arbitrary perturbations, the same property holds for $\mathcal{F}|_{\Pi_i}$, $\mathcal{S}|_{\Pi_i}$, and $\vartheta|_{\Pi_i}$.

Concerning the feasible set mapping, $\mathcal{F}$ is closed and lsc at $\overline{\varphi}$ because $(0_n, \delta)$ is a strong Slater point for $P_{\overline{\varphi}}$ whenever $\delta > \sup_{t \in T} |\overline{\varphi}(t)|$ (take $\varepsilon = \delta - \sup_{t \in T} |\overline{\varphi}(t)|$), so that $\mathcal{F}|_{\Pi_i}$ is closed and, by Theorem 5.1.1, lsc at $\overline{\varphi}$ too. Checking the usc property of $\mathcal{F}|_{\Pi_i}$ at $\overline{\varphi}$ requires ad hoc arguments.

Concerning the optimal set mapping, since $\mathcal{F}$ is lsc at $\overline{\varphi}$, $\mathcal{S}$ is closed at $\overline{\varphi}$ (see Theorem 5.1.4), and it is lsc at $\overline{\varphi}$ if $\mathcal{S}(\overline{\varphi})$ is a singleton set (as it happens if $i = 1$). So, the same statements are true for $\mathcal{S}|_{\Pi_i}$.

Finally, concerning the optimal value function, ϑ is usc at $\overline{\varphi}$ (see Theorem 5.1.5) and so $\vartheta|_{\Pi_i}$ is usc at $\overline{\varphi}$ too. Moreover, if $\mathcal{S}(\overline{\varphi})$ is a nonempty compact set (e.g., when $\overline{\varphi}$ is continuous), then $\vartheta|_{\Pi_i}$ is lsc at $\overline{\varphi}$, and it is Lipschitz continuous in some neighborhood of $\overline{\varphi}$.

Remark 5.1.1 (Stability of the Feasible Set: Antecedents and Extensions). The characterizations (i)–(vi) and (viii) of the lsc property of $\mathcal{F}$ in Theorem 5.1.1 were given in [103], and (x) in [101]. All of them, together with an equivalent version of (ix), appear in [102, Theorem 6.1 and Exercise 6.5 for (ix)], and (vii) in [115].

Concerning the upper semicontinuity of $\mathcal{F}$, the sufficiency of the boundedness of the feasible set (Corollary 5.1.1) was shown in [104] (also in [102, Corollary 6.2.1]) while the characterization given in Theorem 5.1.3 was proved in [45]. The lsc and usc of $\mathcal{F}$ for systems of infinitely many convex constraints was studied in [179].

Table 5.1 shows the available information on the stability of the feasible set spread on the references listed at the 1st column (they are chronologically ordered, with an asterisk marking those works dealing with systems posed in infinite

Table 5.1 Antecedents on the stability of the feasible set

Ref.	Year	Constr. system	lsc	usc	cl	Top	c	S	T
[202]*	1975	sinf. linear cont.							
[63]	1975	ord. linear						✓	
[123]	1975	sinf.	✓	✓				✓	
[203]*	1976	sinf. $\mathcal{C}^1$							
[222]*	1977	sinf. linear							✓
[29]	1982	sinf. cont.	✓	✓	✓			✓	
[82]	1983	sinf. linear cont.	✓	✓				✓	
[11]*	1983	ord. convex	✓	✓	✓			✓	
[31]	1984	sinf. linear cont.	✓	✓	✓			✓	
[220]	1985	sinf. linear cont.	✓						
[124]	1986	ord. $\mathcal{C}^1$				✓			
[131]	1990	sinf. linear cont.	✓					✓	
[152]	1992	sinf. $\mathcal{C}^1$				✓			
[183]	1994	ord. linear						✓	
[103]	1996	sinf. linear	✓		✓			✓	✓
[101]	1996	sinf. linear				✓			
[104]	1997	sinf. linear			✓				
[151]	1998	sinf. $\mathcal{C}^1$				✓			
[102]	1998	sinf. linear	✓	✓	✓	✓	✓	✓	✓
[177]*	1998	sinf. linear	✓	✓	✓				
[139]	2000	sinf. linear						✓	
[187]*	2000	sinf. linear			✓			✓	
[105]	2001	sinf. linear		✓					
[170]	2001	sinf. convex	✓	✓	✓		✓	✓	
[45]	2002	sinf. linear		✓					
[154]	2004	sinf. linear	✓	✓					
[46]	2005	sinf. linear	✓					✓	
[4]	2006	sinf. linear	✓					✓	
[3]	2008	sinf. linear	✓	✓	✓				
[64]	2013	sinf. linear	✓					✓	

dimensional spaces). For the sake of brevity we do not include information on the topology defined on the corresponding parameter space. Column 3 informs about the type of constraint system of $\overline{P}$ (there " sinf.," "ord.," and "cont." are abbreviations of "semi-infinite," "ordinary," and "continuous," respectively). The binary information (yes or not) of the columns 4–10 is referred to the following desirable stability properties of the feasible set:

Column 4:   Lower semicontinuity of $\mathcal{F}$ at $\overline{\pi}$ (lsc).
Column 5:   Upper semicontinuity of $\mathcal{F}$ at $\overline{\pi}$ (usc).
Column 6:   Closedness of $\mathcal{F}$ at $\overline{\pi}$ (cl).
Column 7:   Topological stability of $\mathcal{F}$ at $\overline{\pi}$ (top).
Column 8:   Continuity of $\mathcal{F}$ at $\overline{\pi}$ (c).
Column 9:   Slater-type CQ (S).
Column 10:   Tuy stability of $\mathcal{F}$ at $\overline{\pi}$ (T).

Remark 5.1.2 (Hausdorff Semicontinuity of the Feasible Set). Other concepts of lower and upper semicontinuity have been used in [29, 30], and [105] to describe the stability behavior of $\mathcal{F}$:

- $\mathcal{M}$ is *Hausdorff lower semicontinuous* (H-lsc in short) at $\overline{y}$ if for each real number $\varepsilon > 0$ there exists $V \in \mathfrak{N}_{\overline{y}}$, such that $\mathcal{M}(\overline{y}) \subset \mathcal{M}(y) + \varepsilon \mathbb{B}_2$, for all $y \in V$.
- $\mathcal{M}$ is *Hausdorff upper semicontinuous* (H-usc) at $\overline{y}$ if for each real number $\varepsilon > 0$ there exists $V \in \mathfrak{N}_{\overline{y}}$ such that $\mathcal{M}(y) \subset \mathcal{M}(\overline{y}) + \varepsilon \mathbb{B}_2$, for all $y \in V$.

Obviously, Hausdorff lower semicontinuity implies lower semicontinuity and upper semicontinuity implies Hausdorff upper semicontinuity. There is a consensus of experts on the excessive strength of the H-lsc property and the excessive weakness of the H-usc property in our framework.

Remark 5.1.3 (Stability of $\mathcal{F}$ in Problems with Set Constraint and/or Equations). Several papers have been devoted to the qualitative stability analysis of LSIO problems containing linked inequalities to be preserved by any admissible perturbation ([3, 4, 36], where each equation can be interpreted as two zero-sum inequalities), problems including a set constraint (e.g., $x \in \mathbb{R}^n_+$ when the decision variables satisfy physical constraints $x_i \geq 0$, $i = 1, \ldots, n$) or both [3, 4]. For instance, in Example 5.1.2 we could consider fixed the sign constraints, in which case the space of parameters, say Π_1, is a subset of Π. All the properties of $\mathcal{F}$, $\mathcal{S} : \Pi \rightrightarrows \mathbb{R}^n$ and $\vartheta : \Pi \mapsto \overline{\mathbb{R}}$ at $\overline{\pi} \in \Pi_1$ are inherited by $\mathcal{F}_{|\Pi_1}$, $\mathcal{S}_{|\Pi_1}$, and $\vartheta_{|\Pi_1}$, but the converse statements are not necessarily true. So, in this particular case, where $\mathcal{S}$ fails to be lsc at π_3, we should determine whether $\mathcal{S}_{|\Pi_1}$ is lsc or not at π_3 (in fact it is not because arbitrarily small perturbations of c^3 convert the optimal set into a singleton set).

Remark 5.1.4 (Application to Voronoi Cells). Linear semi-infinite systems arise in different branches of mathematics as convex analysis (e.g., the convex and the concave subdifferentials are solutions sets of such type of systems), robust linear complementarity problems [229, Definition 1.10], or computational geometry. The *metric projection* on $T \subset \mathbb{R}^n$, $|T| \geq 2$, is the set-valued mapping $\mathcal{P}_T : \mathbb{R}^n \rightrightarrows \mathbb{R}^n$ associating with each $x \in \mathbb{R}^n$ the set of nearest points in T for the Euclidean distance. So, given $s \in T$, $\mathcal{P}_T^{-1}(s)$ represents the set of all points of $\mathbb{R}^n$ closer to s than to any other element of T. In this framework, the elements of T are called Voronoi sites while $\mathcal{P}_T^{-1}(s)$ is the *Voronoi cell* of s. Until the 1930s, only finite sets of sites were considered, e.g., by Descartes in 1644, Dirichlet in 1850, and Voronoi in 1908, for $n = 2, n = 3$, and $n > 3$, respectively. Delaunay published in 1934 a paper on crystallography where he considered a discrete infinite set T and $n \in \mathbb{N}$. Voronoi cells for finite sets are widely applied in computational geometry, operations research, data compression, economics, marketing, etc. As observed in [226], eliminating $\|x\|_2^2$ in the 2nd inequality of

$$d_2(x, s) \leq d_2(x, t) \Leftrightarrow \|x - s\|_2^2 \leq \|x - t\|_2^2 \text{ for all } t \in T,$$

one gets

$$\mathcal{P}_T^{-1}(s) = \{x \in \mathbb{R}^n : d_2(x,s) \leq d_2(x,t), \, t \in T\}$$
$$= \left\{x \in \mathbb{R}^n : (t-s)'x \leq \frac{\|t\|_2^2 - \|s\|_2^2}{2}, t \in T\right\}.$$

This reformulation has been systematically exploited in [108] and [109] to get geometric information on $\mathcal{P}_T^{-1}(s)$ from the data (T and $s \in T$) and to determine those sets T such that $\mathcal{P}_T^{-1}(s)$ is a given closed convex set, respectively. The effect of different types of perturbations of the nominal data, a couple $(\overline{T}, \overline{s})$ such that $\overline{s} \in \overline{T} \subset \mathbb{R}^n$, on the Voronoi cells has been analyzed in [116].

Remark 5.1.5 (Primal-Dual Stability). In the same way that int dom $\mathcal{F}$ can be interpreted as the set of primal stable consistent parameters (in the sense that sufficiently small perturbations provide primal consistent problems), the topological interior of the main subsets of Π can be seen as the sets of stable parameters in the corresponding sense. Some of these interiors have been characterized in the continuous case [111, 113] and the general case [191], e.g., those corresponding to the partitions (inconsistent-bounded-unbounded or inconsistent-solvable-bounded unsolvable-unbounded) corresponding to the primal problem, the dual problem, or both problems.

Remark 5.1.6 (Stability of the Boundary and the Extreme Points Set). The results on the stability of the boundary summarized in the diagram after Theorem 5.1.3 can be found in [99, 106]. In the latter paper, the relationships between the stability properties of set-valued mappings with closed convex images in $\mathbb{R}^n$ and their corresponding boundary mappings have been analyzed in full generality. The equivalence "$\mathcal{B}$ closed at $\overline{\pi}$ $\Leftrightarrow$ $\mathcal{B}$ usc at $\overline{\pi}$" was used in [94] in order to obtain a sufficient condition for the stable containment between solution sets of linear semi-infinite systems.

The seminal results on the Hausdorff semicontinuity of the set of extreme points of the feasible set of finite systems in [66] were first extended to the feasible set of infinite systems in [100] and then to arbitrary set-valued mappings with convex images in [114].

Remark 5.1.7 (Stability of $\mathcal{S}$ and ϑ : Antecedents and Extensions). Table 5.2 reviews briefly a non-exhaustive list of relevant works, chronologically ordered, on stability of optimization problems of the form

$$\inf f(x, \pi) \text{ s.t. } x \in \mathcal{F}(\pi),$$

where typically,

$$\mathcal{F}(\pi) = \{x \in X : f_t(x) \leq 0, \forall t \in T; \, x \in C\},$$

Table 5.2 Antecedents on the stability of the optimal set

Ref.	Year	X	T	f	$f(t,\cdot)$	$f(\cdot,x)$	K	ϑ	S
[136]	1973	top	abstr	lsc or usc	–	–	–	✓	✓
[29]	1982	$\mathbb{R}^n$	top	lin	fract	arb	$\mathbb{R}^T_-$	✓	✓
[11]	1983	met	abstr	lsc or usc	–	–	–	✓	✓
[30]*	1983	$\mathbb{R}^n$	compH	cont	aff	cont	$\mathbb{R}^T_-$		✓
[82]	1983	$\mathbb{R}^n$	compH	lin	aff	cont	$\mathbb{R}^T_-$		✓
[31]	1984	$\mathbb{R}^n$	compH	lin	aff	cont	$\mathbb{R}^T_-$	✓	✓
[62]	1984	norm	compH	cont	aff	cont	$\mathbb{R}^T_-$	✓	
[157]	1985	$\mathbb{R}^n$	fin	fin conv	fin conv/aff	–	$\mathbb{R}^T_-$	✓	✓
[10]*	1997	met	abstr	lsc	–	–	–		✓
[102]	1998	$\mathbb{R}^n$	arb	lin	aff	arb	$\mathbb{R}^T_-$	✓	✓
[156]	1998	$\mathbb{R}^n$	compH	diff	diff	cont	$\mathbb{R}^T_-$	–	✓
[25]	2000	Ban	arb	cont	cont	–	cl conv	✓	✓
[48]	2001	$\mathbb{R}^n$	arb	lin	aff	arb	$\mathbb{R}^T_-$	✓	✓
[86]	2003	$\mathbb{R}^n$	arb	fin conv	fin conv	arb	$\mathbb{R}^T_-$	✓	✓
[145]	2005	Ban	fin	fin conv	fin conv/aff	–	cl conv	✓	✓
[47]	2005	$\mathbb{R}^n$	arb	lin	aff	arb	$\mathbb{R}^T_-$	✓	✓
[180]*	2006	met	abstr	fin usc	–	–	–		✓
[53]	2006	$\mathbb{R}^n$	arb	lin	aff	arb	$\mathbb{R}^T_-$	✓	
[43]*	2007	$\mathbb{R}^n$	met compH	fin conv	fin conv	cont	$\mathbb{R}^T_-$		✓
[72]*	2007	lcH	arb	lsc conv	lsc conv	arb	$\mathbb{R}^T_-$	✓	
[143]*	2011	$\mathbb{R}^n$	compH	fin conv	fin conv	cont	$\mathbb{R}^T_-$		✓
[73]	2012	Ban	arb	lsc	lsc	arb	$\mathbb{R}^T_-$	✓	✓

X denoting the decision space, C denotes a fixed set constraint C (generally $C = X$), and $\pi \in \Pi$ (the corresponding parameter space). Those works dealing with particular types of perturbations, usually right-hand side (RHS) perturbations, are marked with an asterisk. For comparison purposes, we represent here the functional constraints as $f(t,\cdot) \in K$, where $f(t,x) := f_t(x)$ and K is a given subset of certain partially ordered space Y (e.g., $Y = \mathbb{R}^T$ and $K = \mathbb{R}^T_+$ for our LSIO problem P). We codify the information in the columns 3–8 of Table 5.2 as follows:

Column 3: Banach (Ban), normed (nor), metric (met), locally convex Hausdorff topological vector space (lcH), and topological space (top).

Column 4: finite (fin), arbitrary (arb), and compact Hausdorff topological space (compH). In case of abstract minimization problems (abstr), there is no explicit information on T, $f(t,\cdot)$, $f(\cdot,x)$, and K.

Columns 5–7: affine (aff), linear (lin), fractional (fract), convex (conv), finite valued (fin), continuous (cont), lower semicontinuous (lsc), upper semicontinuous (usc), arbitrary (arb), and continuously differentiable (diff). In case of abstract minimization problems, no direct information on the constraints is available and the usual allowed perturbations are sequential.

Column 8: closed (cl) and convex (conv). For the sake of brevity we do not include in this table information on the parameter space and the only stability concepts considered here are exclusively lower and upper semicontinuity and closedness. This precludes, among other stability concepts related to $\mathcal{S}$, the Lipschitzian and Hölder stabilities [25], the structural stability [151, 153] or the stability of stationary solutions [133].

In [156], T changes with the parameter, but it is always compact and uniformly bounded. Reference [145] deals with the well-posedness of convex programs under linear perturbations of the objective functions and RHS perturbations.

5.2   Stability Restricted to the Domain of the Feasible Set

A new framework has been suggested by the stability theory of zero sum games, where the perturbations are forced to preserve the feasibility of the involved systems [181]. Thus, the challenge consists of characterizing the lsc property of $\mathcal{F}$ restricted to dom $\mathcal{F}$, say $\mathcal{F}^R$, for different types of systems, under perturbations of all the data, of the RHS function, and of the LHS function. Let us mention two results concerning the latter type of perturbations (i.e., when only $\overline{a}$ is perturbed, and $\overline{b}$ and $\overline{c}$ remain fixed), for arbitrary and continuous systems, which constitute a novelty in literature.

Theorem 5.2.1 (Lower Semicontinuity of $\mathcal{F}^R$). *Let $\overline{\pi} = (\overline{c}, \overline{a}, \overline{b})$ be an ordinary LSIO, and consider only arbitrary perturbations of $\overline{a}$. Then the following statements are true*:

(i) *If either $\overline{\pi}$ satisfies SSCQ or $\mathcal{F}(\overline{\pi})$ is a singleton set, then $\mathcal{F}^R$ is lsc at $\overline{\pi}$.*
(ii) *If $\mathcal{F}^R$ is lsc at $\overline{\pi} \in$ int dom $\mathcal{F}$ and $\mathcal{F}(\overline{\pi}) \neq \{0_n\}$, then $\overline{\pi}$ satisfies the SSCQ.*
(iii) *If $\mathcal{F}^R$ is lsc at $\overline{\pi}$ and $\mathcal{F}(\overline{\pi})$ is neither a singleton set nor a subset of a ray, then $\overline{\pi}$ satisfies the SSCQ.*

Theorem 5.2.2 (Dimensional Stability of $\mathcal{F}^R$). *Let $\overline{\pi} = (\overline{c}, \overline{a}, \overline{b}) \in$ dom $\mathcal{F}$ be a continuous LSIO and consider only continuous perturbations of $\overline{a}$. Suppose that $\overline{\pi}$ has not trivial inequalities as constraints and that $0_n \notin \mathcal{F}(\overline{\pi})$. Then, the following statements are equivalent to each other*:

(i) *There exists $V \in \mathfrak{N}_{\overline{\pi}}$ such that dim $\mathcal{F}^R(\pi) =$ dim $\mathcal{F}^R(\overline{\pi})$ for all $\pi \in V$.*
(ii) *SSCQ holds.*
(iii) *dim $\mathcal{F}(\overline{\pi}) = n$.*

Moreover, any of these properties implies that $\mathcal{F}^R$ is lsc at $\overline{\pi}$.

The remarkable difference between Theorem 5.1.1 and Theorems 5.2.1 and 5.2.2 is due to the systems in bd dom $\mathcal{F}$ (as $\mathcal{F}^R$ and $\mathcal{F}$ coincide on int dom $\mathcal{F}$). In the particular case when T is finite, conditions (i)–(iii) in Theorem 5.2.2 hold if and only if one of the following alternatives holds:

(a) $\dim \mathcal{F}(\overline{\pi}) = n$;
(b) $\dim \mathcal{F}(\overline{\pi}) = 0$;
(c) $\mathcal{F}(\overline{\pi})$ is a non-singleton set contained in some open ray (a half-line emanating from 0_n without its apex).

Remark 5.2.1 (Antecedents and Sources). Theorems 5.2.1 and 5.2.2 are [64, Proposition 9] and [64, Proposition 10], respectively.

5.3   Well and Ill-Posedness

The LSIO problem $\overline{P}$ is said to be *well-posed* (*ill-posed*) w.r.t. a certain property when this property is satisfied by any perturbed problem provided that the perturbation is sufficiently small (arbitrarily small perturbations of $\overline{P}$ provide problems satisfying or not that property, respectively). In topological terms, $\overline{P}$ is said to be well-posed (ill-posed) when its associated parameter $\overline{\pi} = (\overline{c}, \overline{a}, \overline{b})$ is an interior point (a boundary point, respectively) of the set of parameters corresponding to problems satisfying such a property. For instance, $\overline{P}$ is well-posed (ill-posed) with respect to primal feasibility when $\overline{\pi} \in \operatorname{int} \operatorname{dom} \mathcal{F}$ ($\overline{\pi} \in \operatorname{bd} \operatorname{dom} \mathcal{F}$, respectively).

The following subspaces of Π are relevant in the primal/dual analysis of well-posedness:

$$
\begin{aligned}
\Pi_c &= \operatorname{dom} \mathcal{F}, & \Pi_c^D &= \operatorname{dom} \mathcal{F}^D, \\
\Pi_i &= \Pi \backslash \Pi_c, & \Pi_i^D &= \Pi \backslash \Pi_c^D, \\
\Pi_s &= \operatorname{dom} \mathcal{S}, & \Pi_s^D &= \operatorname{dom} \mathcal{S}^D, \\
\Pi_b &= \{\pi \in \Pi_c : \vartheta(\pi) > -\infty\}, \\
\Pi_b^D &= \{\pi \in \Pi_c^D : \vartheta^D(\pi) > -\infty\}, \\
\Pi_u &= \Pi_c \backslash \Pi_b, & \Pi_u^D &= \Pi_c^D \backslash \Pi_b^D.
\end{aligned}
$$

Obviously the subspaces Π_i, Π_b, and Π_u (Π_i^D, Π_b^D, and Π_u^D) constitute a partition of Π, called *primal partition* (*dual partition*, respectively).

The following theorem accounts for some results about well-posedness (ill-posedness), confined to the primal setting of general LSIO (not necessarily continuous). Aside the sets $C(\overline{\pi})$, $M(\overline{\pi})$, $N(\overline{\pi})$, and $K(\overline{\pi})$, the statements below involve the following sets associated with $\overline{\pi} = (\overline{c}, \overline{a}, \overline{b})$:

$$
A(\overline{\pi}) := \operatorname{conv}\{\overline{a}_t, \ t \in T\}, \qquad H(\overline{\pi}) := C(\overline{\pi}) + \mathbb{R}_+(0_n, -1),
$$

$$
Z^+(\overline{\pi}) := \operatorname{conv}\{\overline{a}_t, \ t \in T; \ \overline{c}\}, \text{ and } Z^-(\overline{\pi}) := \operatorname{conv}\{\overline{a}_t, \ t \in T; \ -\overline{c}\}.
$$

Obviously, $M(\overline{\pi}) = \mathbb{R}_+ A(\overline{\pi})$.

Theorem 5.3.1 (Primal Well-Posedness).   *Given* $\bar{\pi} = (\bar{c}, \bar{a}, \bar{b}) \in \Pi$, *the following statements hold*:

(i)   *If* $d(\bar{\pi}, \mathrm{bd}\,\Pi_c) < +\infty$, *then* $\bar{\pi} \in \mathrm{int}\,\Pi_i$, $\bar{\pi} \in \mathrm{int}\,\Pi_c$, *or* $\bar{\pi} \in \mathrm{bd}\,\Pi_c$ *if and only if* $0_{n+1} \in \mathrm{int}\,H(\bar{\pi})$, $0_{n+1} \in \mathrm{int}(\mathbb{R}^{n+1}\backslash H(\bar{\pi}))$, *or* $0_{n+1} \in \mathrm{bd}\,H(\bar{\pi})$, *respectively*.

(ii)   *If* $\bar{\pi} \in \mathrm{int}\,\Pi_c$, *then* $\bar{\pi} \in \mathrm{int}(\Pi_c\backslash\Pi_s)$, $\bar{\pi} \in \mathrm{int}\,\Pi_s$, *or* $\bar{\pi} \in \mathrm{bd}\,\Pi_s$ *if and only if* $0_n \in \mathrm{int}(\mathbb{R}^n\backslash Z^-(\bar{\pi}))$, $0_n \in \mathrm{int}\,Z^-(\bar{\pi})$, *or* $0_n \in \mathrm{bd}\,Z^-(\bar{\pi})$, *respectively*.

(iii) $\mathrm{cl}\,\Pi_s = \mathrm{cl}\,\Pi_b$ *and* $\mathrm{int}\,\Pi_s = \mathrm{int}\,\Pi_b$ (*hence* $\mathrm{bd}\,\Pi_s = \mathrm{bd}\,\Pi_b$).

In (i) we are excluding those problems $\bar{\pi}$ such that $d(\bar{\pi}, \mathrm{bd}\,\Pi_c) = +\infty$, problems whose existence and properties are studied in the following chapter. In particular we will see that these problems are inconsistent.

Example 5.3.1.   Consider the problem, in $\mathbb{R}$,

$$\bar{\pi} : \inf_{x \in \mathbb{R}} (-x) \text{ s.t. } 0x \geq 1, \ x \geq -k, \ k \in \mathbb{N}.$$

This problem is obviously inconsistent, but $\bar{\pi} \in \mathrm{bd}\,\Pi_c$ as the problems

$$\pi_\varepsilon : \inf_{x \in \mathbb{R}} (-x) \text{ s.t. } \varepsilon x \geq 1, \ x \geq -k, \ k \in \mathbb{N},$$

with $\varepsilon > 0$ are consistent, and $\pi_\varepsilon \to \bar{\pi}$ as $\varepsilon \downarrow 0$. Therefore, statement (i) above applies (certainly, the reader may verify that $0_2 \in \mathrm{bd}\,H(\bar{\pi})$).

Example 5.3.2.   Let $\bar{\pi}$ be any of the three parameters π_1, π_2 and π_3 considered in Example 1.1.1. We already know that $\bar{\pi} \in \mathrm{int}\,\Pi_c$ and it is easy to check that $0_{n+1} \in \mathrm{int}(\mathbb{R}^{n+1}\backslash H(\bar{\pi}))$; in fact, $d_2(0_{n+1}, H(\bar{\pi})) = (5 + 2\sqrt{2})^{-2}$ (see Example 6.2.1). Since $0_2 \in \mathrm{int}\,A(\bar{\pi}) \subset \mathrm{int}\,Z^-(\bar{\pi})$ we also conclude that $\bar{\pi} \in \mathrm{int}\,\Pi_s$.

It makes sense to call *totally ill-posed problems* to those problems in $(\mathrm{bd}\,\Pi_c) \cap (\mathrm{bd}\,\Pi_s)$, since they are simultaneously ill-posed with respect to both feasibility and solvability. The following characterization of these problems does not involve exclusively the data (so, it is hard to be checked). Let us observe that $\bar{\pi} \in \mathrm{bd}\,\Pi_c$ entails either $\bar{\pi} \in \mathrm{bd}\,\Pi_s$ or $\bar{\pi} \in \mathrm{int}(\Pi\backslash\Pi_s)$.

Theorem 5.3.2 (Total Ill-Posedness).   *If* $\bar{\pi} \in \mathrm{bd}\,\Pi_c$, *then*

$$\bar{\pi} \in \mathrm{bd}\,\Pi_s \Leftrightarrow \text{either } 0_n \in \mathrm{bd}\,Z^+(\bar{\pi}) \text{ or } \bar{\pi} \in \mathrm{cl}(\Pi_c \cap \mathrm{bd}\,\Pi_c). \tag{5.5}$$

Example 5.3.3.   In Example 5.3.1, it is easy to see that in a sufficiently small neighborhood of $\bar{\pi}$, any problem π is either inconsistent or unbounded, i.e., $\bar{\pi} \in \mathrm{int}(\Pi\backslash\Pi_s)$. In fact $0 \in \mathrm{int}\,Z^+(\bar{\pi}) = [-1, 1]$ and $\bar{\pi} \notin \mathrm{cl}(\Pi_c \cap \mathrm{bd}\,\Pi_c)$, since any consistent problem close enough to $\bar{\pi}$ satisfies $0_2 \in \mathrm{int}\left(\mathbb{R}^2\backslash H(\bar{\pi})\right)$, and hence belongs to $\mathrm{int}\,\Pi_c$.

In the continuous setting we have the following primal-dual results:

Theorem 5.3.3 (Primal-Dual Well-Posedness). *In the parametric space of continuous problems Π, the following results hold for $\overline{\pi} = (\overline{c}, \overline{a}, \overline{b}) \in \Pi$:*

(i) $\overline{\pi} \in \mathrm{int}\, \Pi_c \Leftrightarrow \overline{\pi}$ *satisfies the SCQ.*

(ii) $\overline{\pi} \in \mathrm{int}\, \Pi_c^D \Leftrightarrow \overline{c} \in \mathrm{int}\, M(\overline{\pi})$.

(iii) $\overline{\pi} \in \mathrm{int}\, \Pi_b \Leftrightarrow \overline{\pi} \in \mathrm{int}\left(\Pi_b \cap \Pi_b^D\right) \Leftrightarrow \overline{\pi} \in \mathrm{int}\, \Pi_b^D \Leftrightarrow \overline{\pi}$ *satisfies the SCQ and* $c \in \mathrm{int}\, M(\overline{\pi})$.

(iv) $\overline{\pi} \in \mathrm{int}\, \Pi_i \Leftrightarrow \overline{\pi} \in \mathrm{int}\left(\Pi_i \cap \Pi_u^D\right) \Leftrightarrow \overline{\pi} \in \mathrm{int}\, \Pi_i^D \Leftrightarrow (0_n, 1) \in \mathrm{int}\, K(\overline{\pi})$.

(v) $\overline{\pi} \in \mathrm{int}\, \Pi_u^D \Leftrightarrow \overline{\pi} \in \mathrm{int}\left(\Pi_u \cap \Pi_i^D\right) \Leftrightarrow \overline{\pi} \in \mathrm{int}\, \Pi_u \Leftrightarrow \exists y \in \mathbb{R}^n$ *such that* $\overline{c}'y < 0$ *and* $\overline{a}_t' y > 0$ *for all* $t \in T$.

(vi) $\mathrm{int}\left(\Pi_i \cap \Pi_i^D\right) = \mathrm{int}\left(\Pi_b \cap \Pi_i^D\right) = \mathrm{int}\left(\Pi_i \cap \Pi_b^D\right) = \emptyset$.

The condition in (iii), i.e., $\overline{\pi}$ satisfies SCQ and $c \in \mathrm{int}\, M(\overline{\pi})$, also characterizes well-posedness w.r.t. simultaneous boundedness (i.e., $\overline{\pi} \in \mathrm{int}\left(\Pi_b \cap \Pi_b^D\right)$) for a general LSIO, not necessarily continuous. On the contrary, and according to (vi), no continuous LSIO problem is well-posed w.r.t. simultaneous infeasibility, while a general LSIO problem $\overline{\pi}$ is well-posed w.r.t. the same property if and only if

$$0_n \notin \mathrm{cl}\, A(\overline{\pi}), \quad \overline{c} \notin \mathrm{cl}\, M(\overline{\pi}), \quad \text{and} \quad (0_n, 1) \in 0^+ \,\mathrm{cl}\, C(\overline{\pi}).$$

Remark 5.3.1 (Sources and Related Results). Statement (i) in Theorem 5.3.1 comes from Theorems 4 and 5 in [49], (ii) and (iii) are Theorem 2 and Theorem 1 in [51], respectively. Theorem 5.3.2 can be found in [51, Theorem 3], and sufficient conditions for total ill-posedness are established in [54]. In [127], condition (5.5) is characterized in terms of the data, by using the formula for the subdifferential of the supremum function given in [128]. Example 5.3.1 is Example 1 in [51].

In [221] Lipschitz constants for both primal and dual optimal value functions are derived under weaker assumptions of stability, which do not preclude, in all the cases, the existence of duality gap. The allowed perturbations are restricted to the coefficients of the objective function of the corresponding dual problems.

The well-posedness results on continuous LSIO problems given in Theorem 5.3.3 come from [111, 113]. These results have been extended to general LSIO in [191]. Moreover, the well-posedness w.r.t. the existence of a strongly unique optimal solution of the primal problem has been characterized in [115]. The mentioned characterizations of well-posedness, always in terms of the data, may not coincide for both continuous and general LSIO.

Refined results for the case in which Π_b and Π_b^D are split into sets composed by parameters which have compact optimal sets and those for which this desirable property fails are given in [112]. Moreover, in [112] it is also shown that most parameters having either primal or dual bounded associated problems have primal and dual compact optimal sets. This generic property fails for general problems (not continuous), despite almost all the characterizations of the topological interior of above subspaces of Π are still valid for the general LSIO problem [50, 52, 105].

Formulas in terms of the data to compute (or at least to estimate) the distance from a given well-posed problem to ill-posedness w.r.t. primal feasibility, infeasibility, solvability and unsolvability (i.e., the minimum size of those perturbations which provide problems not satisfying the corresponding property) have been provided by Cánovas, Hantoute, López, Parra, and Toledo in a series of papers published from 2005 until 2008 [49–52, 127].

Remark 5.3.2 (Some Open Problems in Qualitative Stability of LSIO Problems).

1. Theorem 5.1.1 and other results above provide a long list of properties which are equivalent to the lower semicontinuity of $\mathcal{F}$ under perturbations of data which include b. It remains to determine, from these properties, which still characterize the lower semicontinuity of $\mathcal{F}$ under perturbations which keep b fixed.
2. Theorem 5.1.3 characterizes the upper semicontinuity of $\mathcal{F}$ under perturbations of the triple, but there are not counterparts of this result for other types of perturbations of $\mathcal{F}$, e.g., those which keep b fixed.

Chapter 6
Quantitative Stability Analysis

6.1 Quantitative Stability of Set-Valued Mappings

The following stability properties of a set-valued mapping are of quantitative nature. Again $\mathcal{M} : Y \rightrightarrows X$ is a set-valued mapping between two spaces Y and X equipped with (possible extended) distances denoted by d.

- $\mathcal{M}$ is *pseudo-Lipschitz* at $(\overline{y}, \overline{x}) \in \mathrm{gph}\,\mathcal{M}$ if there exist $V \in \mathfrak{N}_{\overline{y}}$, $U \in \mathfrak{N}_{\overline{x}}$, and a scalar $\kappa \geq 0$ such that

$$d\,(x, \mathcal{M}\,(y)) \leq \kappa d\,(y, y')\ \ \forall y, y' \in V,\ \forall x \in \mathcal{M}(y') \cap U. \tag{6.1}$$

Observe that this is a stable property, in the sense that if $\mathcal{M}$ is pseudo-Lipschitz at $(\overline{y}, \overline{x})$, it is also pseudo-Lipschitz at points of the graph *around* $(\overline{y}, \overline{x})$. The pseudo-Lipschitz of $\mathcal{M}$ at $(\overline{y}, \overline{x})$ is also known as the *Aubin continuity* or *Lipschitz-like property* of the mapping (see [158, Sect. 1.4] and references therein for details). Moreover, the pseudo-Lipschitz of $\mathcal{M}$ at $(\overline{y}, \overline{x})$ turns out to be equivalent to the *metric regularity* of the inverse $\mathcal{M}^{-1}$ at $(\overline{x}, \overline{y})$; i.e., to the existence of $V \in \mathfrak{N}_{\overline{y}}$, $U \in \mathfrak{N}_{\overline{x}}$, and a scalar $\kappa \geq 0$ such that

$$d\,(x, \mathcal{M}\,(y)) \leq \kappa d\,(y, \mathcal{M}^{-1}\,(x))\ \ \forall y \in V,\ \forall x \in U. \tag{6.2}$$

Indeed, (6.2) trivially implies (6.1), while, assuming (6.1), it can be proved that for smaller neighborhoods we may remove $y' \in V$, requiring only $y' \in \mathcal{M}^{-1}(x)$. Formally, starting from (6.1), one derives (6.2) with the same constant κ and possibly smaller neighborhoods U and V (see again [158, Sect. 1.4]).

The situation is illustrated in Fig. 6.1, where one observes that, according to (6.2), the smaller this bound is, the more stability of the image sets $\mathcal{M}\,(y)$ is held around $\overline{y}$. The infimum of such κ for all the triplets (κ, U, V) verifying (6.2), which does coincide with the infimum of κ for all (κ, U, V) in (6.1), is called *exact Lipschitzian*

M.A. Goberna and M.A. López, *Post-Optimal Analysis in Linear Semi-Infinite Optimization*, SpringerBriefs in Optimization, DOI 10.1007/978-1-4899-8044-1_6, © Miguel A. Goberna, Marco A. López 2014

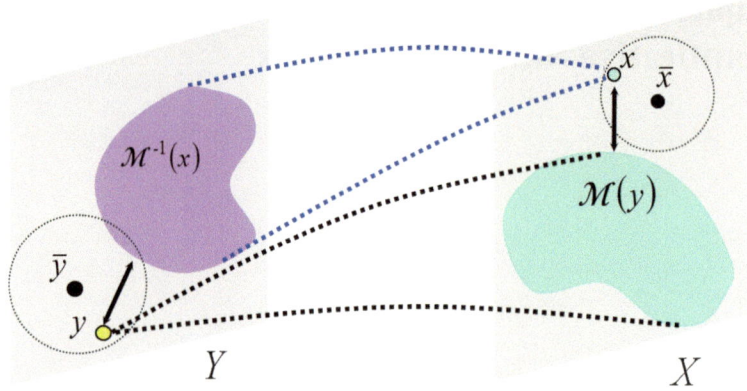

Fig. 6.1 Image sets and pseudo-Lipschitz property

bound (or *Lipschitz modulus*) of $\mathcal{M}$ at $(\overline{y}, \overline{x})$ and it is denoted by lip $\mathcal{M}(\overline{y}, \overline{x})$; so, from (6.1) we easily obtain

$$\text{lip } \mathcal{M}(\overline{y}, \overline{x}) = \limsup_{\substack{y, y' \to \overline{y}, \ y \neq y' \\ x \to \overline{x}, \ x \in \mathcal{M}(y)}} \frac{d(x, \mathcal{M}(y'))}{d(y, y')}.$$

(Observe that y and y' in (6.1) are interchangeable.)

If lip $\mathcal{M}(\overline{y}, \overline{x}) = +\infty$, $\mathcal{M}$ fails to be pseudo-Lipschitz at $(\overline{y}, \overline{x})$.

The pseudo-Lipschitz of $\mathcal{M}$ at $(\overline{y}, \overline{x})$ implies the inner semicontinuity of $\mathcal{M}$ at $(\overline{y}, \overline{x})$ [172, Lemma 2.3]. In fact, the inequality (6.1) for $y' = \overline{y}$ gives

$$d(\overline{x}, \mathcal{M}(y)) \leq \kappa d(y, \overline{y}), \text{ for every } y \in V. \tag{6.3}$$

Then, for V in (6.3) and assuming the nontrivial case $\kappa > 0$, if W is an open set containing $\overline{x}$, we shall take $\eta > 0$ small enough to ensure

$$(x, y) \in W \times V, \text{ whenever } d(x, \overline{x}) < \eta \text{ and } d(y, \overline{y}) < \eta/\kappa.$$

So, appealing to (6.3) one has

$$W \cap \mathcal{M}(y) \neq \emptyset, \text{ whenever } d(y, \overline{y}) < \eta/\kappa,$$

and this accounts for the inner semicontinuity of $\mathcal{M}$ at $(\overline{y}, \overline{x})$.

If both spaces Y and X are normed and $\mathcal{M}$ has closed values, i.e., the images $\mathcal{M}(y)$ are closed sets in X, (6.1) is equivalent (with the same constant κ and the same neighborhoods U and V) to

$$\mathcal{M}(y') \cap U \subset \mathcal{M}(y) + \kappa \|y - y'\| \mathbb{B}_X, \tag{6.4}$$

Fig. 6.2 Lipschitz continuity
of $f(y) = \sqrt{p|y| + (1/16)}$

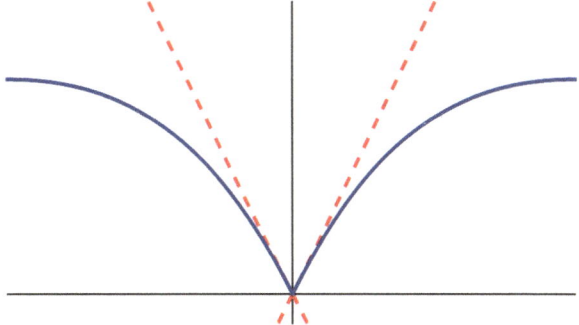

where $\|.\|$ is the norm defined on Y and $\mathbb{B}_X$ is the closed unit ball in X. Without the closedness assumption, (6.4) still implies (6.1), whereas (6.1) implies (6.4) for any $\kappa' > \kappa$.

If $\mathcal{M} \equiv f$ is single valued, (6.1) gives rise to

$$d\left(f(y'), f(y)\right) \leq \kappa d\left(y, y'\right) \ \forall y, y' \in V,$$

in other words, f is *Lipschitz continuous around* $\overline{y}$ (or, equivalently, *strictly continuous at* $\overline{y}$).

Example 6.1.1. In Fig. 6.2 we represent the graph of the function $f : \mathbb{R} \to \mathbb{R}$ defined by

$$f(y) = \sqrt{|y| + (1/16)}$$

which is obviously Lipschitz continuous at any point (why?).

If $f : \mathbb{R}^n \to \mathbb{R}^m$ is of class $\mathcal{C}^1$ on an open set $W \subset \mathbb{R}^n$, then f is Lipschitz-continuous around every $\overline{y} \in W$, and

$$\mathrm{lip}\ f(\overline{y}) = \|\nabla f(\overline{y})\|_2 ,$$

where $\nabla f(\overline{y})$ is the Jacobian matrix of f at $\overline{y}$ (see, for instance, [207, Theorem 8.7]). Difficulties arise whenever f is not differentiable (or not of class $\mathcal{C}^1$). In Fig. 6.2, the reader may verify by a simple geometrical observation that $\mathrm{lip}\ f(0) = 2$.

In Fig. 6.3 we show a mapping $\mathcal{M} : \mathbb{R} \rightrightarrows \mathbb{R}$ which fails to be pseudo-Lipschitz at a particular point $(\overline{y}, \overline{x}) \in \mathrm{gph}\,\mathcal{M}$. This is why, if we approach $\overline{y}$ from the right by points y, y', we see that the ratio $d\left(x, \mathcal{M}(y')\right)/d\left(y, y'\right)$ tends to infinity as the right slope of the upper-boundary of graph $\mathcal{M}$ is $+\infty$.

- $\mathcal{M}$ is *calm* at $(\overline{y}, \overline{x}) \in \mathrm{gph}\,\mathcal{M}$ if there exist $V \in \mathfrak{N}_{\overline{y}}$, $U \in \mathfrak{N}_{\overline{x}}$, and a scalar $\kappa \geq 0$ such that

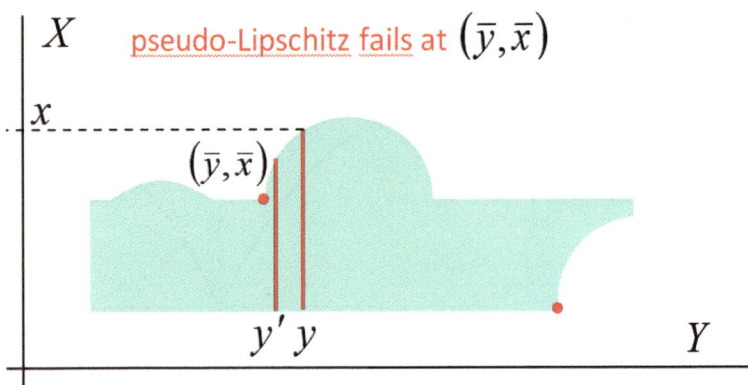

Fig. 6.3 Pseudo-Lipschitz property fails

$$d\left(x, \mathcal{M}\left(\overline{y}\right)\right) \le \kappa d\left(y, \overline{y}\right) \ \forall y \in V, \ \forall x \in \mathcal{M}(y) \cap U. \tag{6.5}$$

Note that $\mathcal{M}(y)$ could be empty for some $y \in V$.

Equivalently, the calmness property of $\mathcal{M}$ can be established in terms of *metric subregularity* of $\mathcal{M}^{-1}$ [75], which reads as the existence of a (possibly smaller) neighborhood U of $\overline{x}$ such that

$$d\left(x, \mathcal{M}\left(\overline{y}\right)\right) \le \kappa d\left(\overline{y}, \mathcal{M}^{-1}\left(x\right)\right), \forall x \in U. \tag{6.6}$$

The constant κ is also known as *local error bound* of $d\left(\overline{y}, \mathcal{M}^{-1}\left(\cdot\right)\right)$ *at* $\overline{x}$, provided that gph $\mathcal{M}$ is locally closed around $(\overline{y}, \overline{x})$.

The infimum of such κ for all the couples (κ, U) verifying (6.6) is called *exact calmness bound* (or *calmness modulus*) of $\mathcal{M}$ at $(\overline{y}, \overline{x})$ and is denoted by clm $\mathcal{M}(\overline{y}, \overline{x})$. From (6.5) we get

$$\operatorname{clm}\mathcal{M}\left(\overline{y}, \overline{x}\right) = \limsup_{\substack{y \to \overline{y}, \ x \to \overline{x} \\ x \in \mathcal{M}(y)}} \frac{d\left(x, \mathcal{M}\left(\overline{y}\right)\right)}{d\left(y, \overline{y}\right)}. \tag{6.7}$$

Observe that (6.6) comes from (6.2) by fixing y at $\overline{y}$, and so, clm $\mathcal{M}(\overline{y}, \overline{x}) \le$ lip $\mathcal{M}(\overline{y}, \overline{x})$. Consequently, pseudo-Lipschitz at $(\overline{y}, \overline{x})$ implies calmness at the same point, but calmness *at* $(\overline{y}, \overline{x})$ does not imply calmness *around* the point.

- $\mathcal{M}$ is said to be *isolatedly calm* (or *locally upper Lipschitz*) if, additionally to (6.6), $\mathcal{M}(\overline{y}) = \{\overline{x}\}$.

If $\mathcal{M} \equiv f$ is single-valued, (6.6) becomes

$$d\left(f\left(y\right), f\left(\overline{y}\right)\right) \le \kappa d(y, \overline{y}), \forall y \in V,$$

Fig. 6.4 Calm, but not
Lipschitz

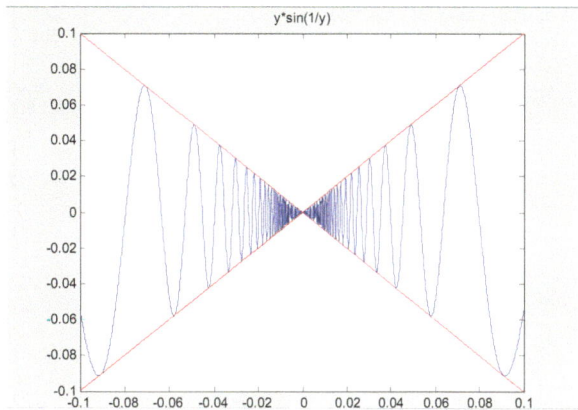

where V is a certain neighborhood of $\overline{y}$. Moreover,

$$\text{clm } f\ (\overline{y}) = \limsup_{y \to \overline{y}} \frac{d\ (f\ (y)\,,\, f\ (\overline{y}))}{d\ (y, \overline{y})}. \tag{6.8}$$

Next we give some examples to compare calmness and Lipschitz continuity.

Example 6.1.2. In Fig. 6.4 we represent the graph of the function

$$f(y) = \begin{cases} y \sin \frac{1}{y}, & \text{if } y \ne 0, \\ 0, & \text{if } y = 0. \end{cases}$$

We have, from (6.8)

$$\text{clm } f\ (0) = \limsup_{y \to 0} \frac{d\ (f\ (y)\,,\, f\ (0))}{d\ (y, 0)} = \limsup_{y \to 0} \left| \sin \frac{1}{y} \right| = 1.$$

(Take the sequence $y_k = 1/(2k\pi + (\pi/2)),\ k = 1, 2, \ldots$)

On the other hand, if we take a new sequence $y'_k = 1/(2k\pi + (3\pi/2))$, $k = 1, 2, \ldots$, also converging to 0, we can write

$$\text{lip } f\ (0) = \limsup_{y,y' \to 0} \frac{d\ (f\ (y)\,,\, f\ (y'))}{d\ (y, y')} \ge \lim_{k \to \infty} \frac{d\ (f\ (y_k)\,,\, f\ (y'_k))}{\left| y_k - y'_k \right|}$$

$$= \lim_{k \to \infty} (4k + 2) = +\infty,$$

i.e., $\text{lip } f\ (0) = +\infty$, entailing that f is calm at 0 but it is not Lipschitz continuous at this point.

Fig. 6.5 Calm and Lipschitz
at $y = 0$

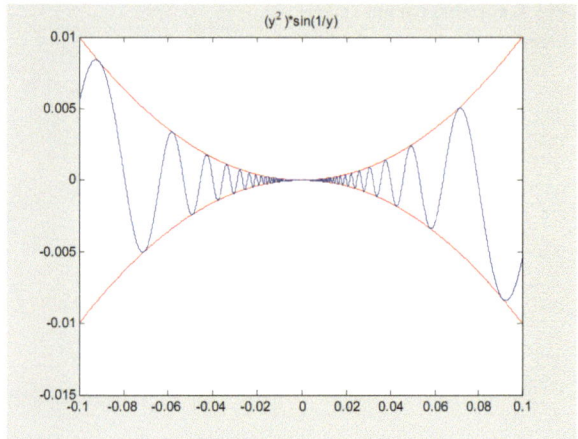

Example 6.1.3. The function in Fig. 6.5

$$f(y) = \begin{cases} y^2 \sin \frac{1}{y}, & \text{if } y \neq 0, \\ 0, & \text{if } y = 0, \end{cases}$$

is not of class $\mathcal{C}^1$ at 0 because

$$f'(y) = \begin{cases} 2y \sin \frac{1}{y} - \cos \frac{1}{y}, & \text{if } y \neq 0, \\ 0, & \text{if } y = 0, \end{cases}$$

and $\lim_{y \to 0} f'(y)$ does not exist.

The reader will verify easily that clm $f(0) = 0$. In order to calculate lip $f(0)$ we make the following considerations. For $y \neq y'$, the mean-value theorem (it does not require that $f \in \mathcal{C}^1$) yields the existence of $z \in]y, y'[$ such that

$$\frac{|f(y') - f(y)|}{|y' - y|} = |f'(z)|,$$

and, for every $\varepsilon > 0$, with y and y' close enough to 0 we can ensure that

$$\frac{|f(y') - f(y)|}{|y' - y|} = |f'(z)|$$

$$\leq |2z \sin(1/z) - \cos(1/z)| \leq |2z \sin(1/z)| + |\cos(1/z)| \leq \varepsilon + 1.$$

(Remember that $f'(0) = 0$.) Therefore,

$$\text{lip } f(0) = \limsup_{y, y' \to 0} \frac{|f(y') - f(y)|}{|y' - y|} \leq 1.$$

Moreover, if

$$y_k = \frac{1}{2k\pi}, \qquad k = 1, 2, \ldots,$$

we have $f'(y_k) = -1$, and by the definition of derivative, there will exist, for each k, a real number y_k' such that

$$\left| \frac{|f(y_k') - f(y_k)|}{|y_k' - y_k|} - |f'(y_k)| \right| = \left| \frac{|f(y_k') - f(y_k)|}{|y_k' - y_k|} - 1 \right| \leq \frac{1}{k}.$$

Taking limits above, for $k \to \infty$,

$$\lim_{k\to\infty} \frac{|f(y_k') - f(y_k)|}{|y_k' - y_k|} = 1,$$

and so lip $f(0) = 1$.

In this chapter some results on the pseudo-Lipschitz property and the calmness of $\mathcal{F}$ and $\mathcal{S}$ are established, and estimates of the corresponding modulus are also provided.

- A *global error bound* of $d\left(\overline{y}, \mathcal{M}^{-1}(\cdot)\right)$, provided that gph $\mathcal{M}$ is closed, is any nonnegative scalar $\kappa < +\infty$ such that

$$d\left(x, \mathcal{M}\left(\overline{y}\right)\right) \leq \kappa d\left(\overline{y}, \mathcal{M}^{-1}(x)\right), \text{ for all } x \in X. \qquad (6.9)$$

If such a scalar κ exists, then the following finite supremum (which coincides with the infimum of the κ's satisfying (6.9)) constitutes a kind of *condition rate* for $\overline{\pi}$:

$$0 \leq \tau(\mathcal{M}, \overline{y}) := \sup_{x\in X} \frac{d\left(x, \mathcal{M}\left(\overline{y}\right)\right)}{d\left(\overline{y}, \mathcal{M}^{-1}(x)\right)},$$

under the convention $0/0 = 0$.

6.2   Quantitative Stability of the Feasible Set Mapping

The first quantitative stability criterion considered in this section is the distance to ill-posedness. This concept was introduced, in the context of conic linear systems, by Renegar in [200]. Besides constituting itself a quantitative measure of the stability (and well-posedness) of the system, this distance becomes a key tool in the analysis of the complexity of certain algorithms, as interior point methods in [201], the ellipsoid algorithm in [84], or a generalization of von Neumann's algorithm studied in [78]. In different subsections we give expressions for $d(\pi, \mathrm{bd\, dom}\,\mathcal{F})$, $d(\pi, \mathrm{bd\, dom}\,\mathcal{S})$, etc.

6.2.1 Distance to Ill-Posedness with Respect to Consistency

Concerning the estimation of distances to ill-posedness, in a series of papers ([49–52], etc.) various formulas were provided to translate the calculus of pseudodistances from $\overline{\pi} = (\overline{c}, \overline{a}, \overline{b})$ to the sets of ill-posed problems into the calculus of distances from the origin to a suitable set in the Euclidean space ($\mathbb{R}^n$ or $\mathbb{R}^{n+1}$). We consider in this subsection arbitrary norms in $\mathbb{R}^n$ and $\mathbb{R}^{n+1}$, and their associated extended distances d in the parameter space Π given by (2.4). In [49, Sect. 5] it is proved that

$$d\,(\overline{\pi}, \mathrm{bd\, dom}\,\mathcal{F}) = \left| \sup_{x \in \mathbb{R}^n} \inf_{t \in T} \frac{\overline{a}'_t x - \overline{b}_t}{\|(x, -1)\|_*} \right|.$$

In the same paper an alternative expression for $d\,(\overline{\pi}, \mathrm{bd\, dom}\,\mathcal{F})$ is given, with a clear geometrical meaning. A first difficulty we face here is that, in our framework, it could happen that $d\,(\overline{\pi}, \mathrm{bd\, dom}\,\mathcal{F}) = \infty$. For instance, if the constraint system of $\overline{\pi}$ is $\overline{\sigma} := \{\frac{1}{r} x \geq r, \ r \in \mathbb{N}\}$, one can easily check that

$$d\,(\overline{\pi}, \mathrm{bd\, dom}\,\mathcal{F}) = \left| \sup_{x \in \mathbb{R}} \inf_{r \in \mathbb{N}} \frac{(1/r)x - r}{\|(x, -1)\|_*} \right| = |-\infty| = \infty.$$

This situation is only possible when the set $C(\overline{\pi})$ is unbounded. Observe also that such a system $\overline{\pi}$ must be inconsistent. Otherwise, i.e., if $\overline{\pi} \in \mathrm{dom}\,\mathcal{F}$, we can obtain an inconsistent system from it just by replacing an arbitrarily chosen constraint of $\overline{\pi}$, for instance $\overline{a}'_{t_0} x \geq \overline{b}_{t_0}$ for a certain $t_0 \in T$, by $0'_n x \geq 1$, and so

$$d\,(\overline{\pi}, \mathrm{bd\, dom}\,\mathcal{F}) = d\,(\overline{\pi}, \mathrm{bd}(\Pi \setminus \mathrm{dom}\,\mathcal{F}))$$
$$= d\,(\overline{\pi}, \Pi \setminus \mathrm{dom}\,\mathcal{F}) \leq \left\| (\overline{a}_{t_0}, \overline{b}_{t_0}) - (0_n, 1) \right\|,$$

the second equality coming from [49, Corollary 1].

Aside the marginal function $\overline{g}$, we also make use of the function $\overline{f} : \mathbb{R}^{n+1} \to \mathbb{R} \cup \{+\infty\}$, given by

$$\overline{f}(x, \lambda) := \sup\{\overline{b}_t \lambda + \overline{a}'_t x : t \in T\}. \tag{6.10}$$

Obviously, $\overline{f}(-x, 1) = \overline{g}(x)$. The function $\overline{f}$ is the support function of $\mathrm{cl}\, C(\overline{\pi})$ and it is a lower semicontinuous sublinear function, whose effective domain satisfies [134, Proposition V.2.2.4]

$$\mathrm{cl}(\mathrm{dom}\,\overline{f}) = (0^+\,\mathrm{cl}\,C(\overline{\pi}))^\circ.$$

The subdifferential of $\overline{f}$ at $(x,\lambda) \in \mathrm{dom}\,\overline{f}$ is

$$\partial\overline{f}(x,\lambda) = \{(u,\mu) \in \mathrm{cl}\,C(\overline{\pi}):\ \overline{f}(x,\lambda) = \langle u,x\rangle + \mu\lambda\}.$$

In particular

$$\partial\overline{f}(0_{n+1}) = \mathrm{cl}\,C(\overline{\pi}).$$

Given $\pi_r = (c^r, a^r, b^r)$ and defining

$$f_r(x,\lambda) := \sup\{b_t^r\lambda + (a_t^r)'x : t \in T\},$$

one has for every $(x,\lambda) \in \mathbb{R}^{n+1}$ and $r = 1,2,\ldots$

$$\left|\overline{f}(x,\lambda) - f_r(x,\lambda)\right| \le d(\pi_r,\overline{\pi})\|(x,\lambda)\|_*, \tag{6.11}$$

and $d(\pi_r,\overline{\pi}) < +\infty,\ r = 1,2,\ldots,$ will imply $\mathrm{dom}\,\overline{f} = \mathrm{dom}\,f_r$. Moreover, if $\lim_{r\to\infty} d(\pi_r,\overline{\pi}) = 0$, the sequence $f_r, r = 1,2,\ldots,$ will converge to $\overline{f}$ pointwisely.

The following theorem provides a couple of characterizations of these abnormal systems.

Theorem 6.2.1 (Systems Such That $d_\infty(\overline{\pi}, \mathbf{bd\,dom}\,\mathcal{F}) = +\infty$). *The following statements are equivalent:*

(i) $d_\infty(\overline{\pi}, \mathrm{bd\,dom}\,\mathcal{F}) = +\infty$;

(ii) the marginal function of $\overline{\pi}$, $\overline{g}(x) := \sup_{t\in T}\left\{\overline{b}_t - \overline{a}_t'x\right\}$, is improper, i.e.,
  $\overline{g} = +\infty$;

(iii) $(0_n, 1) \in 0^+\,\mathrm{cl}\,C(\overline{\pi})$.

Let us sketch the proof of the equivalence of (i) and (ii). Suppose first that $d_\infty(\overline{\pi}, \mathrm{bd\,dom}\,\mathcal{F}) = +\infty$ and, reasoning by contradiction, assume that $\overline{g}(x_0) < +\infty$. Then, if we consider the perturbed system $\pi = (\overline{c}, \overline{a}, \overline{b} + b)$, where b is the constant function $b_t = -\overline{g}(x_0)$, it is obvious that $\pi \in \mathrm{dom}\,\mathcal{F}$ (as $x_0 \in \mathcal{F}(\pi)$), and we get the contradiction $d_\infty(\overline{\pi}, \mathrm{bd\,dom}\,\mathcal{F}) \le d_\infty(\overline{\pi},\pi) = |\overline{g}(x_0)| < +\infty$.

On the other hand, if $\overline{g} = +\infty$ and we consider $\pi = (c,a,b)$ such that $d_\infty(\overline{\pi},\pi)$ is finite, and we define

$$f(x,\lambda) := \sup\{b_t\lambda + a_t'x : t \in T\},$$

then (6.11) yields for any $x \in \mathbb{R}^n$

$$g(x) = f(-x, 1) \geq \overline{f}(-x, 1) - d_\infty (\overline{\pi}, \pi) \|(x, \lambda)\|_1 = \overline{g}(x) - d_\infty (\overline{\pi}, \pi) \|(x, \lambda)\|_1$$

and $g = +\infty$, entailing $\pi \notin \operatorname{dom} \mathcal{F}$.

To show the equivalence of (ii) and (iii) we use the function $\overline{f}$ defined in (6.10). Assume that (iii) holds, that is, for every fixed $(u, \mu) \in \operatorname{cl} C(\overline{\pi})$ and all $\gamma \geq 0$, we have

$$(u, \mu) + \gamma (0_n, 1) \in \operatorname{cl} C(\overline{\pi}) = \partial \overline{f}(0_{n+1}).$$

Thus, for each $x \in \mathbb{R}^n$,

$$\overline{f}(-x, 1) \geq ((u, \mu) + \gamma (0_n, 1))'(-x, 1) = \mu + \gamma - \langle u, x \rangle,$$

for all $\gamma \geq 0$, i.e., $\overline{g}(x) = \overline{f}(-x, 1) = +\infty$.

Conversely, assume that (ii) holds but $(0_n, 1) \notin 0^+ \operatorname{cl} C(\overline{\pi})$. By the separation theorem, there will exist $(v, \alpha) \in \mathbb{R}^{n+1} \setminus \{0_{n+1}\}$ and a scalar β such that

$$(v, \alpha)'(z, \mu) \leq \beta < \alpha \text{ for all } (z, \mu) \in 0^+ \operatorname{cl} C(\overline{\pi}).$$

Since $0^+ \operatorname{cl} C(\overline{\pi})$ is a closed convex cone, we see from the previous inequalities that $\beta = 0, \alpha > 0$, and

$$(v, \alpha) \in [0^+ \operatorname{cl} C(\overline{\pi})]^\circ = \operatorname{cl}(\operatorname{dom} \overline{f}).$$

Consequently, applying Theorem 6.1 in [205], and taking into account that $\overline{f}$ is positively homogeneous, there would exist $\overline{x}$ satisfying $\overline{f}(-\overline{x}, 1) = \overline{g}(\overline{x}) < +\infty$, and this contradicts (ii).

In Sect. 3.1 we introduced the hypographical set $H(\overline{\pi}) := C(\overline{\pi}) + \mathbb{R}_+ \{(0_n, -1)\}$. Observe that

$$H(\overline{\pi}) = \mathbb{R}^{n+1} \Rightarrow d_\infty (\overline{\pi}, \operatorname{bd} \operatorname{dom} \mathcal{F}) = +\infty. \tag{6.12}$$

Actually this is a straightforward consequence of Theorem 6.2.1 and the fact that

$$\overline{g}(x) = \sup_{(a,b) \in H(\overline{\pi})} \{b - a'x\}.$$

Theorem 6.2.2 (Distance to Ill-Posedness w.r.t. Consistency). *If $d(\overline{\pi}, \operatorname{bd} \operatorname{dom} \mathcal{F}) < +\infty$, we have*

$$d(\overline{\pi}, \operatorname{bd} \operatorname{dom} \mathcal{F}) = d(0_{n+1}, \operatorname{bd} H(\overline{\pi})). \tag{6.13}$$

Equation (6.13) is valid for an arbitrary norm in $\mathbb{R}^{n+1}$, $\|\cdot\|$, and the associated distance $d(\cdot, \cdot)$ in the parameter space (see (2.4)).

Example 6.2.1. We are revisiting Example 1.1.1. Any problem $\overline{\pi}$ considered there (i.e., for any $\overline{c} \in \mathbb{R}^n$) belongs to int dom $\mathcal{F}$ (SCQ holds) and the reader may verify that

$$d_2\left(\overline{\pi}, \text{bd dom} \, \mathcal{F}\right) = (5 + 2\sqrt{2})^{-1/2},$$

and that the perturbed problem $\pi = (\overline{c}, a, b)$ where

$$(a_t, b_t) := (\overline{a}_t, \overline{b}_t) - u, \ t \in T,$$

with

$$u = (5 + 2\sqrt{2})^{-1}(1, 1, -1 - \sqrt{2}),$$

is such that $\pi \in \text{bd dom} \, \mathcal{F}$ and $d_2\left(\overline{\pi}, \pi\right) = d_2\left(\overline{\pi}, \text{bd dom} \, \mathcal{F}\right)$. Observe that, in fact,

$$\frac{1 + \sqrt{2}}{5 + 2\sqrt{2}}(a_{\pi/4}, b_{\pi/4}) + \frac{4 + \sqrt{2}}{2(5 + 2\sqrt{2})}((a_2, b_2) + (a_3, b_3)) = 0_3,$$

and $0_3 \in \text{bd} \, H(\pi)$.

6.2.2   Pseudo-Lipschitz Property of $\mathcal{F}$

If we translate the corresponding definition into our context, it turns out by (6.2) that $\mathcal{F} : \Pi \rightrightarrows X$ is *pseudo-Lipschitz* at $(\overline{\pi}, \overline{x}) \in \text{gph} \, \mathcal{F}$ if there exist $V \in \mathfrak{N}_{\overline{\pi}}$, $U \in \mathfrak{N}_{\overline{x}}$, and a scalar $\kappa \geq 0$ such that

$$d_2\left(x, \mathcal{F}(\pi)\right) \leq \kappa d_\infty\left(\pi, \mathcal{F}^{-1}(x)\right) \ \forall \pi \in V, \forall x \in U, \qquad (6.14)$$

and the associated *Lipschitz modulus* is

$$\text{lip} \, \mathcal{F}(\overline{\pi}, \overline{x}) = \limsup_{(x, \pi) \to (\overline{x}, \overline{\pi})} \frac{d_2\left(x, \mathcal{F}(\pi)\right)}{d_\infty\left(\pi, \mathcal{F}^{-1}(x)\right)},$$

under the convention $0/0 = 0$.

This property has important consequences in the overall stability of the constraint system of π, as well as in its sensitivity analysis, and affects even the numerical complexity of the algorithms conceived for finding a solution of the system. Many authors explored the relationship of this property with standard constraint quali- fications as Mangasarian–Fromovitz CQ, SCQ, Robinson CQ, etc. For instance, in [156] the relationships among metric regularity, metric regularity with respect to right-hand side perturbations, and the extended Mangasarian–Fromovitz CQ are established in a non-convex differentiable framework.

Equation (6.14) means that the distance $d_2(x, \mathcal{F}(\pi))$ is bounded from above by $\kappa d_\infty(\pi, \mathcal{F}^{-1}(x))$ around $(\overline{\pi}, \overline{x})$, and this fact is especially useful because $d_\infty(\pi, \mathcal{F}^{-1}(x))$ is a kind of residual, easily computable in general.

To illustrate this difference, let us consider the simplest model of a continuous nominal problem $\overline{\pi} = (\overline{c}, \overline{a}, \overline{b})$ with continuous perturbations of the right-hand side function $\overline{b}$, i.e., we are considering only parameters of the form $\pi = (\overline{c}, \overline{a}, \overline{b} + b)$, with associated constraint system

$$\{\overline{a}'_t x \geq \overline{b}_t + b_t, \ t \in T\},$$

where perturbations b belong to $\mathcal{C}(T)$. In this case the residual accounts for the supremum of constraint violations of π by x, i.e.,

$$d_\infty(\pi, \mathcal{F}^{-1}(x)) = \max_{t \in T} \left[\overline{b}_t + b_t - \overline{a}'_t x \right]_+ ,$$

whereas to calculate $d_2(x, \mathcal{F}(\pi)) = \inf_{x' \in \mathcal{F}(\pi)} \|x - x'\|_2$ is much more complicated. In fact, if $\overline{\pi}$ has a Slater point, we get the extended Ascoli formula [58]

$$d_2(x, \mathcal{F}(\pi)) = \max_{(a,\alpha) \in C(\pi)} \frac{[\alpha - a'x]_+}{\|a\|_2},$$

with $[\cdot]_+$ denoting the positive part, and $C(\pi) := \mathrm{conv}\{(\overline{a}_t, \overline{b}_t + b_t) : t \in T\}$.

Let us sketch the proof of this result as it comes by a straightforward application of our standard tools. In fact we prove that, for every $x \in \mathbb{R}^n$ we have

$$d_2(x, \mathcal{F}(\pi)) = \max_{(a,\alpha) \in C(\pi)} d_2(x, H(a,\alpha)) = \max_{(a,\alpha) \in C(\pi)} \frac{[\alpha - a'x]_+}{\|a\|_2},$$

where $H(a,\alpha) := \{x \in \mathbb{R}^n : a'x \geq \alpha\}$. The proof has different steps:

(1) $d_2(x, H(a,\alpha)) = \frac{[\alpha - a'x]_+}{\|a\|_2}$ is the well-known Ascoli formula.

(2) The inequality "$\geq$", i.e., $d_2(x, \mathcal{F}(\pi)) \geq \sup_{(a,\alpha) \in C(\pi)} d_2(x, H(a,\alpha))$ follows from the fact that $\mathcal{F}(\pi) \subset H(u,v)$ for every $(u,v) \in C(\pi)$.

(3) The converse inequality "$\leq$" follows from the following argument. In the nontrivial case, i.e., $x \notin \mathcal{F}(\pi)$, consider its orthogonal projection $\hat{x} \in \mathcal{F}(\pi)$. Then, $u := (x - \hat{x}) / \|x - \hat{x}\|_2$ satisfies

$$d_2(x, \mathcal{F}(\pi)) = \|x - \hat{x}\|_2 = u'(x - \hat{x}),$$

and

$$u'(z - \hat{x}) \geq 0 \text{ for all } z \in \mathcal{F}(\pi).$$

(4) Since the constraint system of π is Farkas–Minkowski, by Farkas Lemma there will exist $\lambda \in \mathbb{R}_+^{(T)}$ such that

$$\begin{pmatrix} u \\ u'\hat{x} \end{pmatrix} = \sum_{t \in T} \lambda_t \begin{pmatrix} \overline{a}_t \\ \overline{b}_t + b_t \end{pmatrix}.$$

Then, defining $\gamma := \sum_{t \in T} \lambda_t \; (> 0, \text{ since } u \neq 0_n)$, and

$$\begin{pmatrix} \overline{a} \\ \overline{\alpha} \end{pmatrix} := \gamma^{-1} \begin{pmatrix} u \\ u'\hat{x} \end{pmatrix} = \sum_{t \in T} \frac{\lambda_t}{\gamma} \begin{pmatrix} \overline{a}_t \\ \overline{b}_t + b_t \end{pmatrix} \in C(\pi),$$

one has

$$d_2(x, \mathcal{F}(\pi)) = u'(\hat{x} - x) = \frac{\overline{\alpha} - (\overline{a})'x}{\|\overline{a}\|_2} \leq \max_{(a,\alpha) \in C(\pi)} \frac{[\alpha - a'x]_+}{\|a\|_2}.$$

The following theorem characterizes the pseudo-Lipschitz property of $\mathcal{F}$, under continuous perturbations of $\overline{b}$, and gives an explicit formula for the Lipschitz modulus.

Theorem 6.2.3 (Pseudo-Lipschitz Property of $\mathcal{F}$, Canonical Perturbations).
Let $\overline{\pi} = (\overline{c}, \overline{a}, \overline{b})$ be a continuous LSIO and consider only continuous perturbations of $\overline{b}$ ($\overline{c}$ and $\overline{a}$ remain fixed). Then, the following statements hold:

(i) $\mathcal{F}$ is pseudo-Lipschitz at $(\overline{\pi}, \overline{x})$ for every $\overline{x} \in \mathcal{F}(\overline{\pi})$ if and only if $\overline{\pi}$ satisfies SCQ.

(ii) Assume that $\mathcal{F}$ is pseudo-Lipschitz at $(\overline{\pi}, \overline{x})$. Then:

(ii-1) If $\overline{x}$ is a Slater point for $\overline{\pi}$, then $\operatorname{lip} \mathcal{F}(\overline{\pi}, \overline{x}) = 0$.

(ii-2) If $\overline{x}$ is not a Slater point for $\overline{\pi}$, then

$$\operatorname{lip} \mathcal{F}(\overline{\pi}, \overline{x}) = \max \left\{ \frac{1}{\|a\|_2} : \begin{pmatrix} a \\ a'\overline{x} \end{pmatrix} \in C(\overline{\pi}) \right\} > 0. \tag{6.15}$$

Thanks to statement (i) and Theorem 5.1.1, together with the compactness of $C(\overline{\pi})$, $\mathcal{F}$ is pseudo-Lipschitz at $(\overline{\pi}, \overline{x})$ for every $\overline{x} \in \mathcal{F}(\overline{\pi})$ if and only if $0_{n+1} \notin C(\overline{\pi})$, and this happens if and only if $\mathcal{F}$ is lsc at $\overline{\pi}$.

If we consider the marginal function associated with the nominal problem $\overline{\pi}$, $\overline{g}(x) := \sup_{t \in T}\{\overline{b}_t - \overline{a}'_t x\}$, it turns out that $\overline{g}(x) > 0$ if and only if $x \notin \mathcal{F}(\overline{\pi})$. If $(\overline{\pi}, \overline{x}) \in \text{gph}\,\mathcal{F}$ with $\overline{g}(\overline{x}) = 0$, we know that

$$\partial \overline{g}(\overline{x}) = \text{conv}\left\{-\overline{a}_t : \overline{b}_t - \overline{a}'_t \overline{x} = 0,\ t \in T\right\} = \text{conv}\left\{-\overline{a}_t : t \in T(\overline{x})\right\},$$

and if $\mathcal{F}$ is pseudo-Lipschitz at $(\overline{\pi}, \overline{x}) \in \text{gph}\,\mathcal{F}$ with $\overline{g}(\overline{x}) = 0$, (6.15) can be written as

$$\text{lip}\,\mathcal{F}(\overline{\pi}, \overline{x}) = \{d_2(0_n, \partial \overline{g}(\overline{x}))\}^{-1}.$$

Example 6.2.2. We revisit again Example 1.1.1, and the three (optimal) points analyzed there:

(a) If $x^1 = 0_2$, then $\{a \in \mathbb{R}^2 : (a, a'x^1) = (a, 0) \in C(\overline{\pi})\} = \text{conv}\{(1, 0), (0, 1)\}$, and $\text{lip}\,\mathcal{F}(\overline{\pi}, x^1) = \sqrt{2}$.

(b) With $x^2 = (1/\sqrt{2})(1, 1)$ we get $\{a \in \mathbb{R}^2 : (a, a'x^2) = \left(a, (1/\sqrt{2})(a_1 + a_2)\right)$ $\in C(\overline{\pi})\} = \{-(1/\sqrt{2})(1, 1)\}$, so that $\text{lip}\,\mathcal{F}(\overline{\pi}, x^2) = 1$.

(c) Finally, with $x^3 = (0, 1)$, $\{a \in \mathbb{R}^2 : (a, a'x^3) = (a_1, a_2, a_2) \in C(\overline{\pi})\} = \{(0, -1)\}$ and $\text{lip}\,\mathcal{F}(\overline{\pi}, x^3) = 1$.

A remarkably more involved case arises when perturbations of all the coefficients are allowed. The following theorem deals with this case.

Theorem 6.2.4 (Pseudo-Lipschitz Property of $\mathcal{F}$, General Perturbations). *Let $\overline{\pi} = (\overline{c}, \overline{a}, \overline{b})$ be an ordinary LSIO, and consider arbitrary perturbations of $\overline{a}$ and $\overline{b}$. Then, the following statements hold:*

(i) *$\mathcal{F}$ is pseudo-Lipschitz at $(\overline{\pi}, \overline{x})$ for every $\overline{x} \in \mathcal{F}(\overline{\pi})$ if and only if $\overline{\pi}$ satisfies the SSCQ (or it satisfies any other condition equivalent to the lsc of $\mathcal{F}$ at $\overline{\pi}$).*

(ii) *Assume that $\mathcal{F}$ is pseudo-Lipschitz at $(\overline{\pi}, \overline{x})$ and that the set $\{\overline{a}_t,\ t \in T\}$ is bounded, then :*

   (a) *If $\overline{x}$ is a SS point for $\overline{\pi}$, then $\text{lip}\,\mathcal{F}(\overline{\pi}, \overline{x}) = 0$.*
   (b) *If $\overline{x}$ is not a SS point for $\overline{\pi}$, then*

$$\text{lip}\,\mathcal{F}(\overline{\pi}, \overline{x}) = \|(\overline{x}, -1)\|_2 \max\left\{\frac{1}{\|a\|_2} : \begin{pmatrix} a \\ a'\overline{x} \end{pmatrix} \in \text{cl}\,C(\overline{\pi})\right\} > 0. \tag{6.16}$$

Again Theorem 5.1.1 yields the following equivalence: $\mathcal{F}$ is pseudo-Lipschitz at $(\overline{\pi}, \overline{x})$ for every $\overline{x} \in \mathcal{F}(\overline{\pi})$ if and only if $0_{n+1} \notin \text{cl}\,C(\overline{\pi})$, if and only if $\mathcal{F}$ is lsc at $\overline{\pi}$.

The maximum in (6.16) is attained and it is positive. Observe that $\text{gph}\,\mathcal{F}$ is not convex and, therefore, standard tools in variational analysis as the Robinson–Ursescu theorem (see, for instance, [75]) do not apply here.

6.2.3   Calmness of $\mathcal{F}$

Let $\overline{\pi} = (\overline{c}, \overline{a}, \overline{b})$ be a LSIO and consider again only perturbations of $\overline{b}$ ($\overline{c}$ and $\overline{a}$ remain fixed). The same function $\overline{g}$ can also be used for analyzing the calmness of $\mathcal{F}$ at $(\overline{\pi}, \overline{x}) \in \mathrm{gph}\,\mathcal{F}$ with $\overline{g}(\overline{x}) = 0$. The calmness of $\mathcal{F}$ at $(\overline{\pi}, \overline{x}) \in \mathrm{gph}\,\mathcal{F}$ reads as the existence of $\kappa \geq 0$ and $U \in \mathfrak{N}_{\overline{x}}$ such that

$$d_2\left(x, \mathcal{F}(\overline{\pi})\right) \leq \kappa d_\infty\left(\overline{\pi}, \mathcal{F}^{-1}(x)\right), \text{ for all } x \in U,$$

where

$$d_\infty\left(\overline{\pi}, \mathcal{F}^{-1}(x)\right) = \sup\left\{\left[\overline{b}_t - \overline{a}'_t x\right]_+, \ t \in T\right\} \tag{6.17}$$

$$= \left[\sup\{\overline{b}_t - \overline{a}'_t x, \ t \in T\}\right]_+$$

$$= [\overline{g}(x)]_+.$$

Next, according to [165, Theorem 1] (and also to [9, Proposition 2.1 and Theorem 5.1]), $\mathcal{F}$ is calm at $(\overline{\pi}, \overline{x}) \in \mathrm{gph}\,\mathcal{F}$ with $\overline{g}(\overline{x}) = 0$ if and only if

$$\liminf_{x \to \overline{x},\ \overline{g}(x) > 0} d_2(0_n, \partial \overline{g}(x)) > 0 \tag{6.18}$$

and

$$\mathrm{clm}\,\mathcal{F}(\overline{\pi}, \overline{x}) = \left\{\liminf_{x \to \overline{x},\ \overline{g}(x) > 0} d_2(0_n, \partial \overline{g}(x))\right\}^{-1}. \tag{6.19}$$

Example 6.2.3. Remember that in Example 1.1.1, for $x^1 = (0, 0)$ we had $\mathrm{lip}\,\mathcal{F}\left(\overline{\pi}, x^1\right) = \sqrt{2}$. Taking the sequence $z^r = (-1/r, -1/r)$, $r = 1, 2, \ldots$, which converges to x^1, we observe $\overline{g}(z^r) = 1/r$ and $\partial \overline{g}(z^r) = \mathrm{conv}\{-(1, 0), -(0, 1)\}$. By (6.19)

$$\mathrm{clm}\,\mathcal{F}(\overline{\pi}, x^1) \geq \left\{\lim_{r \to \infty} d_2(0_n, \partial \overline{g}(z^r))\right\}^{-1} = \sqrt{2} = \mathrm{lip}\,\mathcal{F}(\overline{b}, x^1).$$

Since $\mathrm{clm}\,\mathcal{F}(\overline{\pi}, x^1) \leq \mathrm{lip}\,\mathcal{F}(\overline{\pi}, x^1)$, we conclude that $\mathrm{clm}\,\mathcal{F}(\overline{\pi}, x^1) = \sqrt{2}$.

Now, for $x^3 = (0, 1)$ we had $\mathrm{lip}\,\mathcal{F}\left(\overline{\pi}, x^3\right) = 1$, and if $y^r = (-1/r, 1 - 1/r)$, $r = 2, 3, \ldots$, which obviously converges to x^3, we obtain $\overline{g}(y^r) = 1/r$, $\partial \overline{g}(y^r) = \{-(1, 0)\}$. This time (6.19) yields

$$\mathrm{clm}\,\mathcal{F}(\overline{\pi}, x^3) \geq \left\{\lim_{r \to \infty} d_2(0_n, \partial \overline{g}(y^r))\right\}^{-1} = 1 = \mathrm{lip}\,\mathcal{F}(\overline{\pi}, x^3),$$

and so $\mathrm{clm}\,\mathcal{F}(\overline{\pi}, x^3) = 1$.

Remark 6.2.1 (Sources and Extensions).
Distance to Ill-Posedness with Respect to Consistency: The equivalence between
(i) and (iii) in Theorem 6.2.1 is Proposition 1 in [49], but the proof given here
of the equivalence between (ii) and (iii) comes from [127, Proposition 1]; and the
equivalence with (ii) is established in [56, Theorem 3]. The implication (6.12)
is given in [52, Proposition 3.2] without appealing to the marginal function $\bar{g}$.
Theorem 6.2.2 is [49, Theorem 6].

Pseudo-Lipschitz Property of the Feasible Set: Theorem 6.2.3 can be found in [34,
Theorem 2.1 and Corollary 3.2], and Theorem 6.2.4 is [37, Theorem 1]. In [156] the
relationships among metric regularity, metric regularity with respect to right-hand
side perturbations, and the extended Mangasarian–Fromovitz CQ are established in
a non-convex differentiable framework. In [140] a local error bound is provided,
under the boundedness of the set $C\left(\bar{\pi}\right)$ but not requiring the existence of a strong
Slater point.

Remark 6.2.2 (Alternative Characterizations of Calmness). Again we consider
only perturbations of $\bar{b}$ in $\bar{\pi} = (\bar{c}, \bar{a}, \bar{b})$.

(a) The following condition also characterizes the calmness of $\mathcal{F}$ at $(\bar{\pi}, \bar{x}) \in \mathrm{gph}\,\mathcal{F}$
with $\bar{g}(\bar{x}) = 0$, and it constitutes a linear version of Theorem 3 in [159]: There
exists $\lambda_0 > 0$ and a neighborhood $U \in \mathfrak{N}_{\bar{x}}$ such that, for each $x \in U$ with
$\bar{g}(x) > 0$ we can find associated $u_x \in \mathbb{R}^n$, $\|u_x\|_2 = 1$, and $\mu_x > 0$ satisfying

$$\bar{g}(x + \mu_x u_x) \leq \bar{g}(x) - \mu_x \lambda_0, \text{ for all } t \in T. \tag{6.20}$$

(b) Let us recall here the well-known *Abadie constraint qualification* (ACQ for
short) at $(\bar{\pi}, \bar{x})$, which in our framework may be written as

$$\mathrm{cl}\, D(\mathcal{F}(\bar{\pi}); \bar{x}) = A\left(\bar{x}\right)^{\circ}. \tag{6.21}$$

Observe that the inclusion $\mathrm{cl}\, D(\mathcal{F}(\bar{\pi}); \bar{x}) \subset A\left(\bar{x}\right)^{\circ}$ always holds (see (1.4)),
and that $\mathrm{cl}\, D(\mathcal{F}(\bar{\pi}); \bar{x})$ is the cone of tangents to $\mathcal{F}(\bar{b})$ at $\bar{x}$. Equation (6.21)
yields to a kind of *asymptotic optimality condition*:

$$\bar{x} \in \mathcal{S}(\bar{\pi}) \Leftrightarrow \bar{c} \in (\mathrm{cl}\, D(\mathcal{F}(\bar{\pi}); \bar{x}))^{\circ} = A\left(\bar{x}\right)^{\circ\circ} = \mathrm{cl}\, A\left(\bar{x}\right).$$

An example, in $\mathbb{R}^3$, where ACQ fails at $(\bar{\pi}, \bar{x})$ is given by

$$\sigma(\bar{\pi}) := \left\{ \begin{array}{r} -(\cos t)\, x_2 - (\sin t)\, x_3 \geq \bar{b}_t, t \in [0, \pi/2], \\ x_3 \geq \bar{b}_t, t = 2, \end{array} \right\}$$

where $T = [0, \pi/2] \cup \{2\}$, $\bar{b}_t = -1$ if $t \in [0, \pi/2]$, $\bar{b}_2 = 1$, and $\bar{x} = (0, 0, 1)'$.
In this case $\mathcal{F}(\bar{\pi}) = \mathbb{R} \times] - \infty, 0] \times \{1\}$, and $\mathrm{cl}\, D(\mathcal{F}(\bar{\pi}); \bar{x}) = \mathbb{R} \times] - \infty, 0] \times \{0\}$
whereas $A\left(\bar{x}\right)^{\circ} = \{a_{\pi/2}, a_2\}^{\circ} = \mathbb{R}^2 \times \{0\}$.

In [57, Theorem 3] it is proved that the fulfilment of ACQ at $(\overline{\pi}, x)$ for $x \in \mathcal{F}(\overline{\pi}) \cap U$, where U is a neighborhood of $\overline{x}$, together with an additional property of uniform boundedness of the scalars generating the unit vectors in $A(\overline{x})$, constitute conjointly a characterization for calmness of $\mathcal{F}$ at $(\overline{\pi}, \overline{x}) \in$ gph $\mathcal{F}$ with $\overline{g}(\overline{x}) = 0$. This characterization is strongly based on [236, Theorem 2.2]. Section 4 in [57] provides an operative formula for computing clm $\mathcal{F}(\overline{\pi}, \overline{x})$ in the case when T is finite.

Remark 6.2.3 (Global Error Bound). In [55, Proposition 4.5] the following global error bound is obtained: Given $\overline{\pi} \in \text{dom } \mathcal{F}$, suppose that there exist K, $\hat{x}$, and $\varepsilon > 0$ such that $\|x\|_2 \leq K$ for all $x \in \mathcal{F}(\overline{\pi})$ and $\overline{a}'_t \hat{x} \geq \overline{b}_t + \varepsilon$ for all $t \in T$. Then $\varepsilon^{-1}(1 + \|\hat{x}\|_2) \max\{1, K\}$ is a global error bound for $\overline{\pi}$. For Example 1.1.1, $\hat{x} = (1/2, 1/2)$ is a Slater point with associated $\varepsilon = 1 - 1/\sqrt{2}$, and $\|x\|_2 \leq 1$ for all $x \in \mathcal{F}(\overline{\pi})$. Hence, $2\sqrt{2} + 3$ is the global error bound given by the expression above.

Remark 6.2.4 (Radius of Metric Regularity). Since the pseudo-Lipschitz property of $\mathcal{F}$ at $(\overline{\pi}, \overline{x})$ is equivalent to the metric regularity of $\mathcal{F}^{-1}$ at $(\overline{x}, \overline{\pi})$, it makes sense to consider the notion of *radius of metric regularity*, which is defined as follows:

$$\text{rad } \mathcal{F}^{-1}(\overline{x}, \overline{\pi}) := \inf_{\ell \in \mathcal{L}} \left\{ \|\ell\| : \begin{array}{c} \mathcal{F}^{-1} + \ell \text{ is not metrically} \\ \text{regular at } (\overline{x}, \overline{\pi}) + \ell(\overline{x}) \end{array} \right\},$$

where $\mathcal{L}$ is the space of linear continuous mappings from $\mathbb{R}^n$ into $\mathcal{C}(T, \mathbb{R})$, i.e., $\ell(x) = \varphi(\cdot)'x$ for a certain $\varphi \in \mathcal{C}(T, \mathbb{R}^n)$. Consequently, if we consider the supremum norm in $\mathcal{C}(T, \mathbb{R})$, one gets

$$\|\ell\| = \sup_{\|x\| \leq 1} \|\ell(x)\|_\infty = \sup_{\|x\| \leq 1} \|\varphi(\cdot)'x\|_\infty$$
$$= \max_{t \in T} \sup_{\|x\| \leq 1} |\varphi(t)'x| = \max_{t \in T} \|\varphi(t)\|_* .$$

Additionally,

$$(\mathcal{F}^{-1} + \ell)(x) = (\overline{a} + \varphi(\cdot))'x - \mathcal{C}_+(T, \mathbb{R}).$$

The following result, called *radius theorem*, is established in [34, Theorem 3.1] (other radius theorems are given in [74, 144, 189]): If $(\overline{\pi}, \overline{x}) \in \text{gph } \mathcal{F}$, with $\overline{\pi} = (\overline{c}, \overline{a}, \overline{b})$ continuous, allowing only continuous perturbations of $\overline{b}$, then:

$$\text{rad } \mathcal{F}^{-1}(\overline{x}, \overline{\pi}) = \frac{1}{\text{lip } \mathcal{F}(\overline{\pi}, \overline{x})}.$$

Remark 6.2.5 (Pseudo-Lipschitz Property of the Boundary Mapping). Larriqueta and Vera de Serio [166] analyze the equivalence between the pseudo-Lipschitz

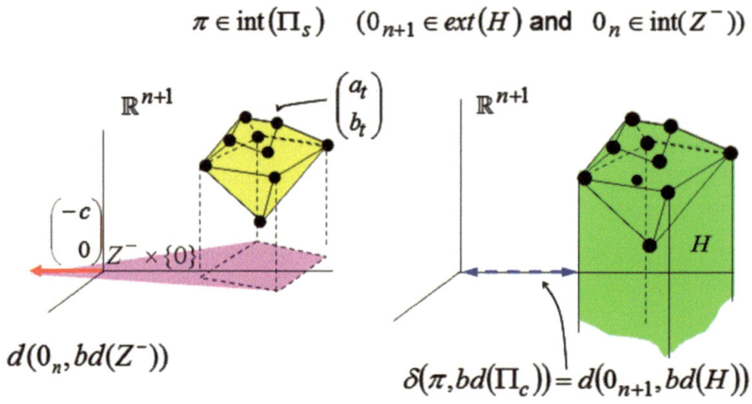

Fig. 6.6 Distance to ill-posedness

property of the boundary set mapping, $\mathcal{B}$, and the stability of the feasible set mapping $\mathcal{F}$ with respect to the consistency. The paper also studies the relationship between the regularity moduli of $\mathcal{B}^{-1}$ and $\mathcal{F}^{-1}$, and provides conditions to assure that the metric regularity of $\mathcal{B}^{-1}$ is equivalent to the lower semi-continuity of $\mathcal{B}$, whose many characterizations are described in Sect. 5.1.

6.3   Quantitative Stability of the Optimal Set Mapping

6.3.1   Distance to the Ill-Posedness with Respect to Solvability

First, in this section, we provide a formula for the distance to the ill-posedness with respect to solvability. For the sake of simplicity, we shall deal only with the Euclidean distance.

Theorem 6.3.1 (Distance to Ill-Posedness w.r.t. Solvability).   *If* $\overline{\pi} = (\overline{c}, \overline{a}, \overline{b}) \in$ cl dom $\mathcal{S}$, *then*

$$d_2(\overline{\pi}, \text{bd dom } \mathcal{S}) = \min\{d_2(0_{n+1}, \text{bd } H\ (\overline{\pi})), d_2(0_n, \text{bd } Z^-\ (\overline{\pi}))\}, \qquad (6.22)$$

where $Z^-\ (\overline{\pi}) := \text{conv}\{\overline{a}_t, t \in T; -\overline{c}\}.$

In Fig. 6.6, ext(H) represents the *exterior* of H, i.e., ext$(H) = \text{int}(\mathbb{R}^{n+1} \backslash H)$. In the example illustrated in this figure, $d_2(0_n, \text{bd } Z^-\ (\overline{\pi})) < d_2(0_{n+1}, \text{bd } H\ (\overline{\pi}))$.

For any possible problem $\overline{\pi}$ considered in Example 1.1.1 (i.e., for any possible $\overline{c}$), $\mathcal{F}(\overline{\pi})$ is bounded as $M\ (\overline{\pi}) = \mathbb{R}^n$. Then, $\mathcal{F}$ is uniformly bounded in some neighborhood of $\overline{\pi}$ according to Lemma 5.1.1. So, $\overline{\pi} \in \operatorname{int} \operatorname{dom} \mathcal{S} \subset \operatorname{cl} \operatorname{dom} \mathcal{S}$. Consequently, theorem above applies and we see that

$$d_2(0_n, \operatorname{bd} Z^-(\overline{\pi})) = (2)^{-1/2} > (5 + 2\sqrt{2})^{-1/2} = d_2(0_{n+1}, \operatorname{bd} H(\overline{\pi})),$$

and $d_2(\overline{\pi}, \operatorname{bd} \operatorname{dom} \mathcal{S}) = (5 + 2\sqrt{2})^{-1/2}$.

Example 6.3.1. Consider the linear programming problem

$$\overline{\pi} : \inf_{x \in \mathbb{R}^2} x_2 \text{ s.t. } -x_1 + x_2 \geq 0, x_1 + x_2 \geq 0, \ (1/4)x_2 \geq 0.$$

Since

$$0_3 \notin H(\overline{\pi}) = \operatorname{conv}\{(-1, 1, 0), (1, 1, 0), (0, 1/4, 0)\} + \mathbb{R}_+\{(0, 0, -1)\}$$

we have and $\overline{\pi} \in \operatorname{int} \operatorname{dom} \mathcal{F}$ by Theorem 5.3.1(i), and moreover

$$0_2 \in \operatorname{int} Z^-(\overline{\pi}) = \operatorname{int} \operatorname{conv}\{(-1, 1), (1, 1), (0, 1/4), -(0, 1)\},$$

entailing, by Theorem 5.3.1(ii), $\overline{\pi} \in \operatorname{int} \operatorname{dom} \mathcal{S}$. Now,

$$1/\sqrt{5} = \|(2/5, -1/5)\|_2 = d_2(0_2, \operatorname{bd} Z^-(\overline{\pi}))$$
$$> d_2(0_3, \operatorname{bd} H(\overline{\pi})) = \|(0, 1/4, 0)\|_2 = 1/4,$$

and

$$d_2(\overline{\pi}_1, \operatorname{bd} \operatorname{dom} \mathcal{S}) = 1/4.$$

A problem $\tilde{\pi} \in \operatorname{bd} \operatorname{dom} \mathcal{S}$ such that $d_2(\overline{\pi}, \operatorname{bd} \operatorname{dom} \mathcal{S}) = d_2(\overline{\pi}, \tilde{\pi})$ is the following

$$\tilde{\pi} : \inf_{x \in \mathbb{R}^2} x_2 \text{ s.t. } \begin{cases} (-1 - 0)x_1 + (1 - 1/4)x_2 \geq 0 - 0, \\ (1 - 0)x_1 + (1 - 1/4)x_2 \geq 0 - 0, \\ (0 - 0)x_1 + (1/4 - 1/4)x_2 \geq 0 - 0. \end{cases}$$

In fact, $\tilde{\pi} \in \operatorname{bd} \operatorname{dom} \mathcal{F}$ as $0_3 \in \operatorname{bd} H(\tilde{\pi})$.

6.3.2   Pseudo-Lipschitz Property of $\mathcal{S}$

From now on we approach the study of $\mathcal{S}$ only in the context of continuous parameters $\overline{\pi} = (\overline{c}, \overline{a}, \overline{b})$, with fixed left-hand side $\overline{a}$, and continuous perturbations of $\overline{c}$ and $\overline{b}$. Let us start by considering the following example:

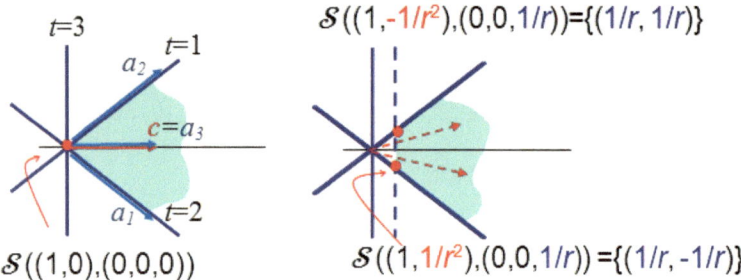

Fig. 6.7 S is not pseudo-Lipschitz

$$\overline{\pi} : \quad \underset{x \in \mathbb{R}^2}{\text{Inf}} \ x_1 \text{ s.t. } x_1 - x_2 \geq 0, \ x_1 + x_2 \geq 0, \ x_1 \geq 0. \qquad (6.23)$$

Here $\overline{\pi} = (\overline{c}, \overline{a}, \overline{b})$, with $\overline{c} = (1, 0)$ and $\overline{b} = (0, 0, 0)$. Obviously, $\mathcal{S}(\overline{\pi}) = \{(0, 0)\}$. In order to see that S is not pseudo-Lipschitz at $(\overline{\pi}, \overline{x})$, consider the perturbed problems (we omit the fixed parameter $\overline{a}$):

$$\pi_r \equiv \left(\left(1, -1/r^2 \right), (0, 0, 1/r) \right), \ \tilde{\pi}_r \equiv \left(\left(1, 1/r^2 \right), (0, 0, 1/r) \right), \ r = 1, 2, \ldots$$

It is straightforward that

$$S(\pi_r) = \{(1/r, 1/r)\}, \quad S(\tilde{\pi}_r) = \{(1/r, -1/r)\}, \ r = 1, 2, \ldots.$$

Therefore

$$d_2(S(\pi_r), S(\tilde{\pi}_r)) = 2/r, \quad \text{while } d_\infty(\pi_r, \tilde{\pi}_r) = 2/r^2, \ r = 1, 2, \ldots,$$

and

$$\text{lip } S(\overline{\pi}, \overline{x}) = \underset{(x,\pi) \to (\overline{x},\overline{\pi})}{\limsup} \frac{d_2(x, S(\pi))}{d_\infty(\pi, (S)^{-1}(x))} \geq \underset{r \to \infty}{\lim} \frac{2/r}{2/r^2} = \infty.$$

See also Fig. 6.7.

In contrast with the situation in the previous example, we show next a problem in which we cannot conclude (see Fig. 6.8) by using the strategy used above, that S does not enjoy the pseudo-Lipschitz property. The new problem is

$$\tilde{\pi} : \quad \underset{x \in \mathbb{R}^2}{\text{Inf}} \ \{x_1 + (1/2)x_2\} \text{ s.t. } x_1 - x_2 \geq 0, \ x_1 + x_2 \geq 0, \ x_1 \geq 0, \qquad (6.24)$$

for which we have again $S(\tilde{\pi}) = \{(0, 0)\}$.

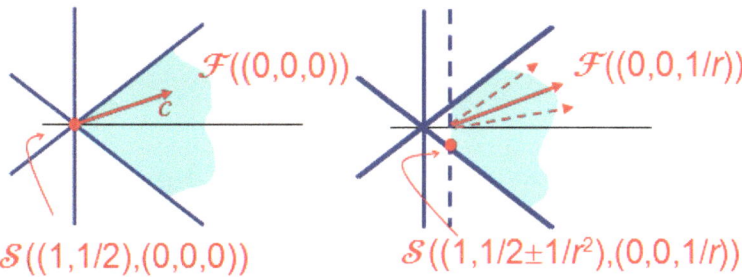

Fig. 6.8 Perturbations do not preclude pseudo-Lipschitz

The following notion is crucial in our approach: given $(\overline{\pi}, \overline{x}) \in \mathrm{gph}\, \mathcal{S}$, the *extended Nürnberger condition* (ENC, in brief) is held at $(\overline{\pi}, \overline{x})$ if $\overline{\pi}$ satisfies the SCQ, and there is no $E \subset T(\overline{x})$ with $|E| < n$ such that $\overline{c} \in \mathrm{cone}\,\{\overline{a}_t,\ t \in E\}$.

The following result gives a complete characterization of the pseudo-Lipschitz property of $\mathcal{S}$ for the LSIO problem in our continuous context. It shows the equivalence among ENC and: (1) the pseudo-Lipschitz property of $\mathcal{S}$ at $(\overline{\pi}, \overline{x})$, (2) the strong Lipschitz stability of $\mathcal{S}$ at $(\overline{\pi}, \overline{x})$ or strong metric regularity of $\mathcal{S}^{-1}$ at $(\overline{x}, \overline{\pi})$ [75, 3G], and (3) a Kojima's type stability under specific perturbations [160]. The fact that the pseudo-Lipschitz property of the optimal set mapping of a parametric optimization problem implies strong Lipschitz stability holds for a rather general class of optimization problems (see, e.g., [158, Corollary 4.7], [75, Theorem 4F.7]).

Theorem 6.3.2 (Pseudo-Lipschitz Property of $\mathcal{S}$ for Continuous LSIO's). *Let us consider the continuous LSIO problem $\overline{\pi} = (\overline{c}, \overline{a}, \overline{b})$ with continuous perturbations of $\overline{c}$ and $\overline{b}$, and $(\overline{\pi}, \overline{x}) \in \mathrm{gph}\, \mathcal{S}$. Then, the following conditions are equivalent:*

(i) $\mathcal{S}$ is pseudo-Lipschitz at $(\overline{\pi}, \overline{x})$.
(ii) $\mathcal{S}$ is single-valued and Lipschitz continuous in some neighborhood of $\overline{\pi}$.
(iii) $\mathcal{S}$ is single-valued and continuous.
(iv) $\mathcal{S}$ is single valued in some neighborhood of $\overline{\pi}$.
(v)   ENC holds at $(\overline{\pi}, \overline{x})$.

Observe that ENC fails for the problem $\overline{\pi}$ in (6.23) but it is held for problem $\tilde{\pi}$ in (6.24), entailing that $\mathcal{S}$ is pseudo-Lipschitz at $(\tilde{\pi}, 0_2)$. In Example 1.1.1, with $\overline{b}$ as it was defined there, $\mathcal{S}$ is pseudo-Lipschitz only at the following pairs:

$$((\overline{c}_1, \overline{c}_2), \overline{b}), \overline{x})\ \text{for}\ \begin{cases} \overline{x} = (1,0) \text{ and } \overline{c}_1 < 0,\ \overline{c}_2 > 0, \\ \overline{x} = (0,1) \text{ and } \overline{c}_1 > 0,\ \overline{c}_2 < 0, \\ \overline{x} = (0,0) \text{ and } \overline{c}_1 > 0,\ \overline{c}_2 > 0. \end{cases}$$

ENC leads us to consider the nonempty family of sets

$$\mathfrak{T}(\bar{x}) := \{E \subset T(\bar{x}) : |E| = n \text{ and } \bar{c} \in \text{cone}\{\bar{a}_t, t \in E\}\}. \tag{6.25}$$

For problem $\tilde{\pi}$ in (6.24), $\mathfrak{T}(0_2) = \{\{1,2\}, \{2,3\}\}$.

It is obvious that, under ENC, if $E \in \mathfrak{T}(\bar{x})$, the vectors $\{\bar{a}_t, t \in E\}$ are linearly independent and so the matrix A_E whose rows are these vectors is non-singular. In [38, Theorem 1] the authors proved that if $\mathcal{S}$ is pseudo-Lipschitz at $(\bar{\pi}, \bar{x}) \in \text{gph } \mathcal{S}$, then

$$\text{lip } \mathcal{S}(\bar{\pi}, \bar{x}) \geq \sup_{E \in \mathfrak{T}(\bar{x})} \left\| A_E^{-1} \right\|.$$

Moreover, the equality holds when T is finite (which solves an open problem proposed in [174]). Here we identify A_E with the endomorphism $x \mapsto A_E x$, with the images space being endowed with the supremum norm $\|\cdot\|_\infty$. Provided that A_E is non-singular, we have

$$\left\| A_E^{-1} \right\| = \max_{\|y\|_\infty \leq 1} \left\| A_E^{-1} y \right\|_2 = \max_{y \in \{-1,1\}^n} \left\| A_E^{-1} y \right\|_2. \tag{6.26}$$

The second equality comes from the fact that $\{-1, 1\}^n$ is the set of extreme points of the closed unit ball corresponding to the norm $\|\cdot\|_\infty$, plus the convexity of the function to be maximized.

For problem $\tilde{\pi}$ in (6.24), lip $\mathcal{S}(\tilde{\pi}, 0_2) = \max\{\left\|(\bar{a}_1, \bar{a}_2)^{-1}\right\|_2, \left\|(\bar{a}_2, \bar{a}_3)^{-1}\right\|_2\} = 5^{1/2}$.

6.3.3 Calmness of $\mathcal{S}$

Again in the context of continuous parameters $\bar{\pi} = (\bar{c}, \bar{a}, \bar{b})$, with continuous perturbations of $\bar{c}$ and $\bar{b}$, calmness of $\mathcal{S}$ is analyzed through the calmness of the *partial solution set mapping* $\mathcal{S}_{\bar{c}} : C(T, \mathbb{R}) \rightrightarrows \mathbb{R}^n$ given by

$$\mathcal{S}_{\bar{c}}(b) := \mathcal{S}(\bar{c}, b),$$

and the so-called *(lower) level set mapping* $\mathcal{L} : \mathbb{R} \times C(T, \mathbb{R}) \rightrightarrows \mathbb{R}^n$ defined by

$$\mathcal{L}(\alpha, b) := \{x \in \mathbb{R}^n \mid \bar{c}'x \leq \alpha; \ \bar{a}'_t x \geq b_t, \ t \in T\}.$$

The calmness of $\mathcal{L}$ at $((\bar{c}'\bar{x}, \bar{b}), \bar{x}) \in \text{gph } \mathcal{L}$ reads as the existence of $\kappa \geq 0$ and a neighborhood $U \in \mathfrak{N}_{\bar{x}}$ such that

$$d_2\left(x, \mathcal{L}(\bar{c}'\bar{x}, \bar{b})\right) \leq \kappa d_\infty\left((\bar{c}'\bar{x}, \bar{b}), \mathcal{L}^{-1}(x)\right), \text{ for all } x \in U,$$

where, obviously, $\mathcal{L}(\overline{c}'\overline{x}, \overline{b}) = \mathcal{S}(\overline{\pi})$, and

$$d_\infty\left((\overline{c}'\overline{x}, \overline{b}), \mathcal{L}^{-1}(x)\right) = \sup\{[\overline{c}'x - \overline{c}'\overline{x}]_+ ; \ [\overline{b}_t - \overline{a}'_t x]_+ , \ t \in T\}$$

$$= \left[\sup\{\overline{c}'x - \overline{c}'\overline{x}; \ \overline{b}_t - \overline{a}'_t x, \ t \in T\}\right]_+ . \qquad (6.27)$$

If we introduce the function $\overline{h} : \mathbb{R}^n \to \mathbb{R}$

$$\overline{h}(x) := \sup\left\{\overline{c}'x - \overline{c}'\overline{x}; \ \overline{b}_t - \overline{a}'_t x, \ t \in T\right\}$$

$$= \max\left\{\overline{c}'x - \overline{c}'\overline{x}, \ \overline{g}(x)\right\}, \qquad (6.28)$$

it turns out that

$$\mathcal{L}(\overline{c}'\overline{x}, \overline{b}) = \left\{x \in \mathbb{R}^n : \overline{h}(x) \le 0\right\} \text{ and } d_\infty\left((\overline{c}'\overline{x}, \overline{b}), \mathcal{L}^{-1}(x)\right) = \left[\overline{h}(x)\right]_+ .$$

Moreover

$$\partial\overline{h}(x) = \begin{cases} \operatorname{conv}\left\{-\overline{a}_t : \overline{b}_t - \overline{a}'_t x = \overline{h}(x), \ t \in T\right\}, & \text{if } \overline{c}'x - \overline{c}'\overline{x} < \overline{h}(x), \\ \operatorname{conv}\left(\left\{-\overline{a}_t : \overline{b}_t - \overline{a}'_t x = \overline{h}(x), \ t \in T\right\} \cup \{\overline{c}\}\right), & \text{if } \overline{c}'x - \overline{c}'\overline{x} = \overline{h}(x). \end{cases}$$

In particular

$$\partial\overline{h}(\overline{x}) = \operatorname{conv}\left(\{-\overline{a}_t : t \in T(\overline{x})\} \cup \{\overline{c}\}\right). \qquad (6.29)$$

As $\overline{a}$ is fixed, and for having a better formulation of the next statements, we shall identify $\overline{\pi}$ with $(\overline{c}, \overline{b})$. Then, if $\left((\overline{c}, \overline{b}), \overline{x}\right) \in \operatorname{gph} \mathcal{S}$ satisfies SCQ, Proposition 6.3.1 below establishes the equivalence among the calmness of $\mathcal{S}$ at $\left((\overline{c}, \overline{b}), \overline{x}\right)$ and the calmness of $\mathcal{L}$ at $\left((\overline{c}'\overline{x}, \overline{b}), \overline{x}\right)$, allowing us to use the characterization for calmness of the set mapping given in (6.18):

Proposition 6.3.1 (Calmness of $\mathcal{S}$). *If $\left((\overline{c}, \overline{b}), \overline{x}\right) \in \operatorname{gph} \mathcal{S}$, and we assume that SCQ holds, then the following statements are equivalent:*

(i) $\mathcal{S}$ is calm at $\left((\overline{c}, \overline{b}), \overline{x}\right)$;

(ii) $\mathcal{S}_{\overline{c}}$ is calm at $(\overline{b}, \overline{x})$;

(iii) $\mathcal{L}$ is calm at $\left((\overline{c}'\overline{x}, \overline{b}), \overline{x}\right)$;

(iv) $\liminf_{x \to \overline{x}, \ \overline{h}(x)>0} d_2(0_n, \partial\overline{h}(x)) > 0$.

Now we provide a theorem characterizing the isolated calmness of the argmin mapping:

Theorem 6.3.3 (Isolated Calmness of S). *Let $\overline{\pi} = (\overline{c}, \overline{a}, \overline{b})$, with $\overline{c} \neq 0_n$, be a continuous LSIO and consider only continuous perturbations of $\overline{c}$ and $\overline{b}$. If $S(\overline{\pi}) = \{\overline{x}\}$ and SCQ holds, then the following assertions are equivalent:*

(i) $\overline{c} \in \operatorname{int} A(\overline{x})$;

(ii) $\overline{x}$ *is a strongly unique minimizer of* $\overline{\pi}$;

(iii) S *is isolatedly calm at* $((\overline{c}, \overline{b}), \overline{x})$;

(iv) $S_{\overline{c}}$ *is isolatedly calm at* $(\overline{b}, \overline{x})$.

If T is finite, the conditions above are equivalent to the uniqueness of $\overline{x}$ as optimal solution of $\overline{\pi}$.

As the authors of [42] pointed out, the equivalence (i)$\Leftrightarrow$(iii) can also be obtained from Theorem 6.3.1. Thanks to (6.29), $\overline{c} \in \operatorname{int} A(\overline{x})$ is equivalent to

$$0_n \in \operatorname{int}(\operatorname{conv}(\{-\overline{a}_t : t \in T(\overline{x})\} \cup \{\overline{c}\})) = \operatorname{int} \partial \overline{h}(\overline{x}).$$

Let $\kappa > 0$ be such that $\kappa^{-1} \mathbb{B}_2 \subset \partial \overline{h}(\overline{x})$. Then, for every $x \in \mathbb{R}^n$ we have

$$\langle u, x - \overline{x} \rangle \leq \kappa \overline{h}(x) \text{ for all } u \in \mathbb{B}_2,$$

and, if $x \neq \overline{x}$, taking $u = (x - \overline{x})/\|x - \overline{x}\|_2$ we get

$$d_2(x, S(\overline{c}, \overline{b})) = \|x - \overline{x}\|_2 \leq \kappa \overline{h}(x);$$

hence, taking into account that $S(\overline{c}, \overline{b}) = \mathcal{L}(\overline{c}' x, \overline{b})$, Theorem 6.3.1 yields the aimed calmness property. The same argument (applied backwards) also shows the necessity of $\overline{c} \in \operatorname{int} A(\overline{x})$. We call the attention of the reader on the fact that we have actually proved that (i) is equivalent to the existence of a global error bound of $\overline{h}$ at $\overline{x}$.

In the rest of this section we approach the problem of estimating $\operatorname{clm} S((\overline{c}, \overline{b}), \overline{x})$. More precisely, we shall provide a lower bound for this modulus under SCQ and uniqueness of the optimal solution. It turns out that this lower bound is also upper bound in linear programming, and therefore, it represents the exact modulus in linear programming under uniqueness of the optimal solution. In addition, we shall give a new upper bound which has the virtue of relying only on the nominal data.

Remember that we are considering all the time a continuous LSIO $\overline{\pi} = (\overline{c}, \overline{a}, \overline{b})$, with continuous perturbations of $\overline{c}$ and $\overline{b}$, and assume that SCQ holds. Associated with $\overline{x} \in S(\overline{c}, \overline{b})$ we introduce the family of active index sets:

$$\mathfrak{K}(\overline{x}) := \{E \subset T(\overline{x}) : |E| \leq n \text{ and } \overline{c} \in \operatorname{cone}\{a_t, \ t \in E\}\}.$$

$\mathfrak{K}(\overline{x}) \neq \emptyset$ by Theorem 1.1.3, and it relates to the different KKT representations of vector $\overline{c}$, after applying Carathéodory Lemma.

Fig. 6.9 Interpretation of the
function f_D

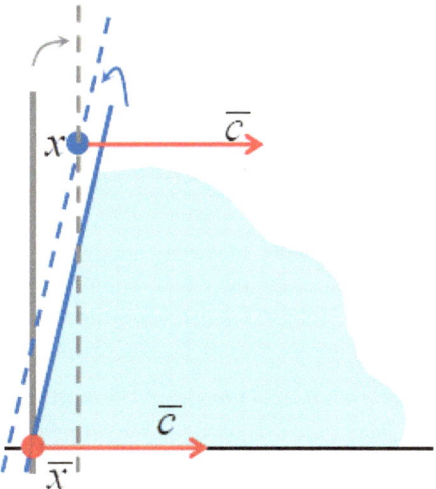

Associated with each $E \in \Re(\overline{x})$ we consider the supremum function

$$f_E(x) := \sup \left\{ \overline{b}_t - \overline{a}_t' x, \ t \in T; \ -(\overline{b}_t - \overline{a}_t' x), \ t \in E \right\}$$

$$= \sup \left\{ \overline{b}_t - \overline{a}_t' x, \ t \in T \setminus E; \ \left| \overline{a}_t' x - \overline{b}_t \right|, \ t \in E \right\}.$$

Obviously $f_E(x) \geq 0$ for all $x \in \mathbb{R}^n$, and $f_E(x)$ is the smallest perturbation size on $\overline{b}$ to make x be a KKT point (hence optimal), with E as an associated KKT set.

Let us illustrate the meaning of the function f_E with a simple example:

$$\pi : \ \text{Inf}_{x \in \mathbb{R}^2} \{ x_1 \ \text{s.t.} \ x_1 \geq 0, \ x_1 - (1/4)x_2 \geq 0, \ x_2 \geq 0 \}.$$

Here $\overline{b} = 0_3$ and the only optimal solution is $\overline{x} = 0_2$. For $E = \{1\}$ and $x = (1, 6)$ we see in Fig. 6.9 that

$$f_{\{1\}}(1, 6) := \sup\{-1, -1 + (1/4)6, -6, -(-1)\} = 1.$$

Only constraints 1 and 2 need to be perturbed in order to get optimality at x with $\overline{c} \in \text{cone}\{a_1\}$ (and the perturbation size is $f_{\{1\}}(x) = 1$).

Let us give some further information about the functions $f_E, E \in \Re(\overline{x})$:

1. $f_E : \mathbb{R}^n \to \mathbb{R}_+$ is a convex function.
2. $[f_E = 0] \subset \mathcal{S}(\overline{c}, \overline{b})$ for all $E \in \Re(\overline{x})$.
3. Under uniqueness, i.e., if $\mathcal{S}(\overline{c}, \overline{b}) = \{\overline{x}\}$, we have $[f_E = 0] = \mathcal{S}(\overline{c}, \overline{b})$ for all $E \in \Re(\overline{x})$, and one can also check that, for $x \notin \mathcal{S}(\overline{c}, \overline{b})$,

$$d_2(x, \overline{x}) \equiv d_2(x, \mathcal{S}(\overline{c}, \overline{b})) \geq \frac{f_E(x)}{d_2(0_n, \partial f_E(x))}.$$

The following result is established in [44, Theorem 5].

Theorem 6.3.4 (Lower Bound for the Calmness Modulus of $\mathcal{S}$). *Consider the continuous LSIO $\bar{\pi} = (\bar{c}, \bar{a}, \bar{b})$, with continuous perturbations of $\bar{c}$ and $\bar{b}$. If we assume that SCQ holds and that $S(\bar{c}, \bar{b}) = \{\bar{x}\}$, then*

$$\mathrm{clm}\, \mathcal{S}((\bar{c}, \bar{b}), \bar{x}) \geq \mathrm{clm}\, \mathcal{S}_{\bar{c}}(\bar{b}, \bar{x}) \geq \sup_{E \in \mathfrak{K}(\bar{x})} \limsup_{\substack{x \to \bar{x} \\ f_E(x) > 0}} \frac{1}{d_2\left(0_n, \partial f_E(x)\right)}.$$

We consider next the important particular case of linear programming. The following theorem [44, Theorem 6 and Corollary 1] shows that the lower bound above for $\mathrm{clm}\, \mathcal{S}\left((\bar{c}, \bar{b}), \bar{x}\right)$ is also an upper bound, without requiring neither SCQ nor the uniqueness of $\bar{x}$ as an optimal solution of $\bar{\pi}$.

Theorem 6.3.5 (1st Upper Bound for the Calmness Modulus of $\mathcal{S}$ in LO). *Assume that T is finite. If $\bar{x} \in S(\bar{c}, \bar{b})$, then*

$$\mathrm{clm}\, \mathcal{S}((\bar{c}, \bar{b}), \bar{x}) = \mathrm{clm}\, \mathcal{S}_{\bar{c}}(\bar{b}, \bar{x}) \leq \sup_{E \in \mathfrak{K}(\bar{x})} \limsup_{\substack{x \to \bar{x} \\ f_E(x) > 0}} \frac{1}{d_2\left(0_n, \partial f_E(x)\right)}.$$

If, in addition, $S(\bar{c}, \bar{b}) = \{\bar{x}\}$, the inequality above becomes an equality.

The following step is addressed to provide an upper bound for the calmness modulus in linear programming relaying only on problem's data. Recall that in linear programming, according to Theorem 6.3.3,

$$S(\bar{c}, \bar{b}) = \{\bar{x}\} \Leftrightarrow \bar{c} \in \mathrm{int}\, A\,(\bar{x}),$$

which entails the existence of some subset of indices $E \subset T(\bar{x})$, with $|E| = n$, such that

$$\bar{c} \in \mathrm{cone}\{\bar{a}_t, \ t \in E\} \text{ and } A_E := (\bar{a}_t)_{t \in E} \text{ is non-singular.}$$

In other words, the following family of active index sets is nonempty:

$$\mathfrak{G}(\bar{x}) := \{E \in \mathfrak{T}(\bar{x}) : \ A_E \text{ non-singular}\},$$

where the family $\mathfrak{T}(\bar{x})$ has been introduced in (6.25). In fact, under the ENC condition, and thanks to Carathéodory Lemma, both families coincide, i.e., $\mathfrak{G}(\bar{x}) = \mathfrak{T}(\bar{x})$. In [44, Theorem 8] the following upper bound is given:

Theorem 6.3.6 (2nd Upper Bound for the Calmness Modulus of $\mathcal{S}$ in LO). *If T is finite and $\bar{c} \in \mathrm{int}\, A(\bar{x})$, then*

$$\mathrm{clm}\, \mathcal{S}((\bar{c}, \bar{b}), \bar{x}) \leq \max_{E \in \mathfrak{G}(\bar{x})} \left\| A_E^{-1} \right\|. \tag{6.30}$$

Table 6.1 Calmness moduli of $\mathcal{S}$ in LO and LSIO

	LSIO under SCQ	LO
Not requiring uniqueness	Open problem	$\text{clm}\,\mathcal{S}_{\overline{c}}(\overline{b},\overline{x})$ $= \text{clm}\,\mathcal{S}((\overline{c},\overline{b}),\overline{x})$ $\leq C1$
$\mathcal{S}(\overline{c},\overline{b}) = \{\overline{x}\}$	$C1 \leq \text{clm}\,\mathcal{S}_{\overline{c}}(\overline{b},\overline{x})$ $\leq \text{clm}\,\mathcal{S}((\overline{c},\overline{b}),\overline{x})$	$\text{clm}\,\mathcal{S}_{\overline{c}}(\overline{b},\overline{x})$ $= \text{clm}\,\mathcal{S}((\overline{c},\overline{b}),\overline{x})$ $= C1 \leq C2$

Fig. 6.10 Second upper bound attained (a)

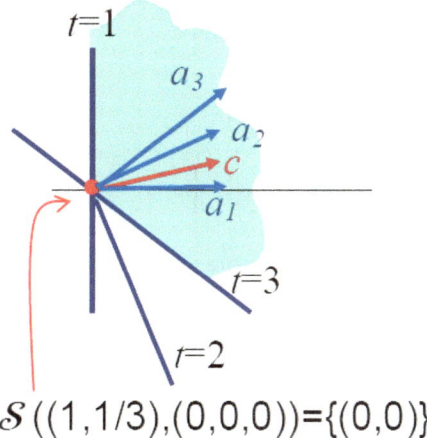

$$\mathcal{S}\left((1,1/3),(0,0,0)\right)=\{(0,0)\}$$

Table 6.1, where

$$C1 := \sup_{\substack{E\in\mathfrak{K}(\overline{x})}} \limsup_{\substack{x\to\overline{x}\\ f_E(x)>0}} \frac{1}{d_2\left(0,\partial f_E\left(x\right)\right)} \quad\text{and}\quad C2 := \max_{E\in\mathfrak{G}(\overline{x})} \left\|A_E^{-1}\right\|,$$

summarizes the results in this subsection (Fig. 6.10).

Example 6.3.2 (2nd Upper Bound Attained).   Consider the problem

$$\pi : \text{Inf}_{x\in\mathbb{R}^2} \{x_1 + (1/3)x_2 \mid x_1 \geq 0,\ x_1 + (1/2)x_2 \geq 0,\ x_1 + x_2 \geq 0\}.$$

We easily verify that $\mathfrak{G}(\overline{x}) = \{\{1,2\},\{1,3\}\}$, $\left\|A_{\{1,2\}}^{-1}\right\| = \sqrt{17}$, and $\left\|A_{\{1,3\}}^{-1}\right\| = \sqrt{5}$ (according to (6.26). If we consider now the sequence $((b^r,x^r))_{n=1}^{\infty} \subset \text{gph}\,\mathcal{S}_{\overline{c}}$ given by $b^r = (1/r,-1/r,0)$ and $x^r = (-1/r,4/r)$, we have (see also Fig. 6.11)

$$\lim_{r\to+\infty} \frac{\|x^r - \overline{x}\|_2}{\|b^r - \overline{b}\|_{\infty}} = \lim_{r\to+\infty} \frac{\left\|(-\frac{1}{r},\frac{4}{r}) - (0,0)\right\|_2}{\left\|(-\frac{1}{r},\frac{1}{r},0) - (0,0,0)\right\|_{\infty}} = \sqrt{17}.$$

This expression, together with Theorem 6.3.6, entails $\text{clm}\,\mathcal{S}\left((\overline{c},\overline{b}),\overline{x}\right) = \sqrt{17}$.

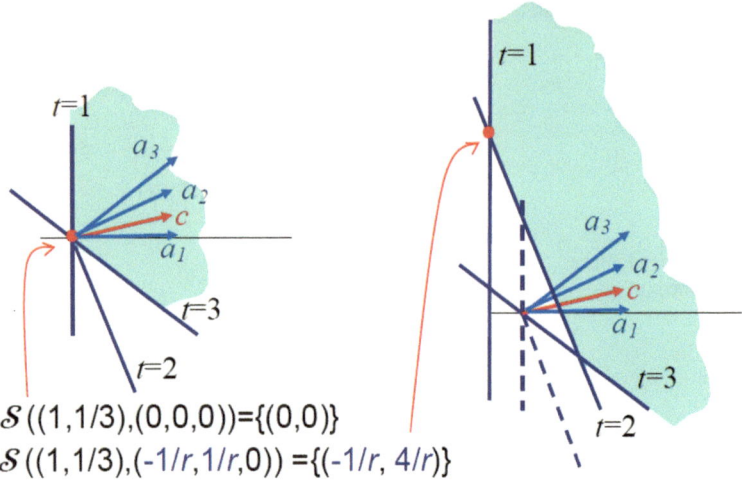

Fig. 6.11 Second upper bound attained (b)

Example 6.3.3 (2nd Upper Bound Not Attained). Consider the problem

$$\pi : \text{Inf}_{x \in \mathbb{R}^2}\{x_1 + (1/3)x_2 \text{ s.t. } x_1 \geq 0, \; x_1 + (1/2)x_2 \geq 0,$$
$$x_1 + x_2 \geq 0, \; x_1 - x_2 \geq 0\}.$$

Now $\mathfrak{G}(\overline{x}) = \{\{1, 2\}, \{1, 3\}, \{2, 4\}, \{3, 4\}\}$ and so

$$\left\|A_{\{1,2\}}^{-1}\right\| = \sqrt{17}, \quad \left\|A_{\{1,3\}}^{-1}\right\| = \sqrt{5}, \quad \left\|A_{\{2,4\}}^{-1}\right\| = \sqrt{17}/3, \quad \left\|A_{\{3,4\}}^{-1}\right\| = 1.$$

On the other hand, one can check via a non-simple direct calculus (see [44]) that

$$\text{clm}\, \mathcal{S}\left((\overline{c}, \overline{b}), \overline{x}\right) = \sqrt{5} < \left\|A_{\{1,2\}}^{-1}\right\|.$$

Remark 6.3.1 (Sources and Related Results).
Distance to the Ill-Posedness with Respect to Solvability: The results in this subsection, in particular Theorem 6.3.1, are in [50, Theorems 2, 5 and 6].

Pseudo-Lipschitz Property of the Optimal Solution: Theorem 6.3.2 is [43, Theorem 16].

Calmness of the Optimal Solution: Theorem 6.3.1 can be found in [42, Theorem 3.1]. Theorem 6.3.3, characterizing the isolated calmness of the argmin mapping, is Theorem 3 in [35]. The second upper bound in (6.30) equals the Lipschitz modulus when $\mathcal{S}$ is Aubin continuous, according to [37, Corollary 2].

In [39, 40] the following expression for $\mathrm{lip}\,\mathcal{S}\,(\overline{\pi}, \overline{x})$ is established under ENC:

$$\mathrm{lip}\,\mathcal{S}\left((\overline{c}, \overline{b}), \overline{x}\right) = \mathrm{lip}\,\mathcal{S}_{\overline{c}}\left(\overline{b}, \overline{x}\right) = \limsup_{\substack{(z,b)\to(\overline{x},\overline{b}) \\ f_b(z)>0}}\left(d_2(0_n, \hat{\partial} f_b(z))\right)^{-1},$$

where

$$f_b(z) := d_\infty\left(b, (\mathcal{S}_{\overline{c}})^{-1}(z)\right),$$

$(\mathcal{S}_{\overline{c}})^{-1}(z) = \{b \in C(T, \mathbb{R}) : z \in \mathcal{S}(\overline{c}, b)\}$, and $\hat{\partial} f_b(z)$ represents the *Fréchet subdifferential* of f_b at z. Moreover, ENC also allows us to represent f_b in terms of certain *difference of convex (d.c.)* functions, which rely directly on the nominal data [41].

Calmness of $\mathcal{S}$ at $((\overline{c}, \overline{b}), \overline{x})$ always holds when T is finite, even without assuming SCQ at $\overline{b}$. The result follows from [204] (see also [207, Example 9.57] or [75, Theorem 3D.1]) as a consequence of the piecewise polyhedrality of gph $\mathcal{S}$ (i.e., the graph is the finite union of polyhedral sets). Observe that gph $\mathcal{L}$ is polyhedral in the finite case, so that the calmness of $\mathcal{S}$ at $((\overline{c}, \overline{b}), \overline{x})$ can be alternatively derived from Theorem 6.3.1 together with the aforementioned result of Robinson, even without SCQ, since KKT conditions hold for finitely constrained linear problems without SCQ.

Remark 6.3.2 (Two Open Problems in Quantitative Stability of LSIO Problems).

1. Efficient numerical methods for the computation of distances from a given problem to ill-posedness in some of the senses discussed in Sects. 6.2.1 and 6.3.1.
2. Efficient numerical methods for the computation of the Lipschitz and calmness moduli of $\mathcal{F}$ and $\mathcal{S}$, specially for finite T.

References

1. Adler, I., Monteiro, R.: A geometric view of parametric linear programming. Algorithmica **8**, 161–176 (1992)
2. Altinel, I.K., Çekyay, B.Ç., Feyzioğlu, O., Keskin, M.E., Özekici, S.: Mission-based component testing for series systems. Ann. Oper. Res. **186**, 1–22 (2011)
3. Amaya, J., Bosch, P., Goberna, M.A.: Stability of the feasible set mapping of linear systems with an exact constraint set. Set-Valued Anal. **16**, 621–635 (2008)
4. Amaya, J., Goberna, M.A.: Stability of the feasible set of linear systems with an exact constraints set. Math. Methods Oper. Res. **63**, 107–121 (2006)
5. Anderson, E.J., Lewis, A.S.: An extension of the simplex algorithm for semi-infinite linear programming. Math. Program. A **44**, 247–269 (1989)
6. Anderson, E.J., Goberna, M.A., López, M.A.: Simplex-like trajectories on quasi-polyhedral convex sets. Math. Oper. Res. **26**, 147–162 (2001)
7. Auslender, A., Ferrer, A., Goberna, M.A., López, M.A.: Comparative study of RPSALG algorithm for convex semi-infinite programming. Departamento de Estadística e Investigación Operativa, Universidad de Alicante, Spain. Preprint
8. Auslender, A., Goberna, M.A., López, M.A.: Penalty and smoothing methods for convex semi-infinite programming. Math. Oper. Res. **34**, 303–319 (2009)
9. Azé, D., Corvellec, J.-N.: Characterizations of error bounds for lower semicontinuous functions on metric spaces. ESAIM Control Optim. Calc. Var. **10**, 409–425 (2004)
10. Balayadi, A., Sonntag, Y., Zălinescu, C.: Stability of constrained optimization problems. Nonlinear Anal. **28**, 1395–1409 (1997)
11. Bank, B., Guddat, J., Klatte, D., Kummer, B., Tammer, K.: Non-linear Parametric Optimization. Birkhäuser, Basel (1983)
12. Beck, A., Ben-Tal, A.: Duality in robust optimization: primal worst equals dual best. Oper. Res. Lett. **37**, 1–6 (2009)
13. Bellman, R.E., Zadeh, L.A.: Decision making in a fuzzy environment. Manag. Sci. **17**, 141–164 (1970)
14. Ben-Tal, A., El Ghaoui, L., Nemirovski, A.: Robust Optimization. Princeton Series in Applied Mathematics. Princeton University Press, Princeton (2009)
15. Ben-Tal, A., Nemirovski, A.: Robust solutions of linear programming problems contaminated with uncertain data. Math. Program. A **88**, 411–424 (2000)
16. Ben-Tal, A., Nemirovski, A.: Robust optimization—methodology and applications. Math. Program. B **92**, 453–480 (2002)
17. Ben-Tal, A., Teboulle, M.: Expected utility, penalty functions and duality in stochastic nonlinear programming. Manag. Sci. **30**, 1445–1466 (1986)

18. Bennett, K.P., Parrado-Hernández, E.: The interplay of optimization and machine learning research. J. Mach. Learn. Res. **7**, 1265–1281 (2006)
19. Berkelaar, A., Roos, C., Terlaky, T.: The optimal set and optimal partition approach to linear and quadratic programming. In: Gal, T., Greenberg, H. (eds.) Recent Advances in Sensitivity Analysis and Parametric Programming, pp.1–44. Kluwer, Dordrecht (1997)
20. Bertsimas, D., Brown, D.B., Caramanis, C: Theory and applications of robust optimization. SIAM Rev. **53**, 464–501 (2011)
21. Betró, B.: An accelerated central cutting plane algorithm for linear semi-infinite programming. Math. Program. A **101**, 479–495 (2004)
22. Betrò, B.: Numerical treatment of Bayesian robustness problems. Int. J. Approx. Reason. **50**, 279–288 (2009)
23. Betrò, B., Bodini, A.: Generalized moment theory and Bayesian robustness analysis for hierarchical mixture models. Ann. Inst. Stat. Math. **58**, 721–738 (2006)
24. Bhattacharjee, B., Green, W.H., Jr., Barton, P.I.: Interval methods for semiinfinite programs. Comput. Optim. Appl. **30**, 63–93 (2005)
25. Bonnans, J.F., Shapiro, A.: Perturbation Analysis of Optimization Problems. Springer, New York (2000)
26. Boţ, R.I.: Conjugate Duality in Convex Optimization. Springer, Berlin (2010)
27. Boţ, R.I., Jeyakumar, V., Li, G.: Robust duality in parametric convex optimization. Set-Valued Var. Anal. **21**, 177–189 (2013)
28. Box, E.P.: Robustness in the strategy of scientific model building. In: Launer, R.L., Wilkinson, G.N. (eds.) Robustness in Statistics, pp. 201–236. Academic, New York (1979)
29. Brosowski, B.: Parametric Semi-infinite Optimization. Peter Lang, Frankfurt am Main (1982)
30. Brosowski, B.: On the continuity of the optimum set in parametric semi-infinite programming. In: Fiacco, A.V. (ed.) Mathematical Programming with Data Perturbations, pp. 23–48. Marcel Dekker, New York (1983)
31. Brosowski, B.: Parametric semi-infinite linear programming I. Continuity of the feasible set and the optimal value. Math. Program. Study **21**, 18–42 (1984)
32. Cadenas, J.M., Verdegay, J.L.: A primer on fuzzy optimization models and methods. Iran. J. Fuzzy Syst. **3**, 1–22 (2006)
33. Campi, M.C., Garatti, S.: The exact feasibility of randomized solutions of uncertain convex programs. SIAM J. Optim. **19**, 1211–1230 (2008)
34. Cánovas, M.J., Dontchev, A.L., López, M.A., Parra, J.: Metric regularity of semi-infinite constraint systems. Math. Program. B **104**, 329–346 (2005)
35. Cánovas, M.J., Dontchev, A.L., López, M.A., Parra, J.: Isolated calmness of solution mappings in convex semi-infinite optimization. J. Math. Anal. Appl. **350**, 892–837 (2009)
36. Cánovas, M.J., Gómez-Senent, F.J., Parra, J.: Stability of systems of linear equations and inequalities: distance to ill-posedness and metric regularity. Optimization **56**, 1–24 (2007)
37. Cánovas, M.J., Gómez-Senent, F.J., Parra, J.: On the Lipschitz modulus of the argmin mapping in linear semi-infinite optimization. Set-Valued Anal. **16**, 511–538 (2008)
38. Cánovas, M.J., Gómez-Senent, F.J., Parra, J.: Regularity modulus of arbitrarily perturbed linear inequality systems. J. Math. Anal. Appl. **343**, 315–327 (2008)
39. Cánovas, M.J., Hantoute, A., López, M.A., Parra, J.: Lipschitz modulus of the optimal set mapping in convex optimization via minimal subproblems. Pac. J. Optim. **4**, 411–422 (2008)
40. Cánovas, M.J., Hantoute, A., López, M.A., Parra, J.: Stability of indices in KKT conditions and metric regularity in convex semi-infinite optimization. J. Optim. Theory Appl. **139**, 485–500 (2008)
41. Cánovas, M.J., Hantoute, A., López, M.A., Parra, J.: Lipschitz modulus in convex semi-infinite optimization via d.c. functions. ESAIM Control Optim. Calc. Var. **15**, 763–781 (2009)
42. Cánovas, M.J., Hantoute, A., Parra, J., Toledo, J.: Calmness of the argmin mapping in linear semi-infinite optimization. J. Optim. Theory Appl. doi:10.1007/s10957-013-0371-z (in press)
43. Cánovas, M.J., Klatte, D., López, M.A., Parra, J.: Metric regularity in convex semi-infinite optimization under canonical perturbations. SIAM J. Optim. **18**, 717–732 (2007)

44. Cánovas, M.J., Kruger, A.Y., López, M.A., Parra, J., Théra, M.: Calmness modulus of linear semi-infinite programs. SIAM J. Optim., in press
45. Cánovas, M.J., López, M.A., Parra, J.: Upper semicontinuity of the feasible set mapping for linear inequality systems. Set-Valued Anal. **10**, 361–378 (2002)
46. Cánovas, M.J., López, M.A., Parra, J.: Stability of linear inequality systems in a parametric setting. J. Optim. Theory Appl. **125**, 275–297 (2005)
47. Cánovas, M.J., López, M.A., Parra, J.: On the continuity of the optimal value in parametric linear optimization: stable discretization of the Lagrangian dual of nonlinear problems. Set-Valued Anal. **13**, 69–84 (2005)
48. Cánovas, M.J., López, M.A., Parra, J., Todorov, M.I.: Stability and well-posedness in linear semi-infinite programming. SIAM J. Optim. **10**, 82–89 (1999)
49. Cánovas, M.J., López, M.A., Parra, J., Toledo, F.J.: Distance to ill-posedness and the consistency value of linear semi-infinite inequality systems. Math. Program. A **103**, 95–126 (2005)
50. Cánovas, M.J., López, M.A., Parra, J., Toledo, F.J.: Distance to solvability/unsolvability in linear optimization. SIAM J. Optim. **16**, 629–649 (2006)
51. Cánovas, M.J., López, M.A., Parra, J., Toledo, F.J.: Ill-posedness with respect to the solvability in linear optimization. Linear Algebra Appl. **416**, 520–540 (2006)
52. Cánovas, M.J., López, M.A., Parra, J., Toledo, F.J.: Distance to ill-posedness in linear optimization via the Fenchel-Legendre conjugate. J. Optim. Theory Appl. **130**, 173–183 (2006)
53. Cánovas, M.J., López, M.A., Parra, J., Toledo, F.J.: Lipschitz continuity of the optimal value via bounds on the optimal set in linear semi-infinite optimization. Math. Oper. Res. **31**, 478–489 (2006)
54. Cánovas, M.J., López, M.A., Parra, J., Toledo, F.J.: Sufficient conditions for total ill-posedness in linear semi-infinite optimization. Eur. J. Oper. Res. **181**, 1126–1136 (2007)
55. Cánovas, M.J., López, M.A., Parra, J., Toledo, F.J.: Error bounds for the inverse feasible set mapping in linear semi-infinite optimization via a sensitivity dual approach. Optimization **56**, 547–563 (2007)
56. Cánovas, M.J., López, M.A., Parra, J., Toledo, F.J.: Distance to ill-posedness for linear inequality systems under block perturbations: convex and infinite-dimensional cases. Optimization **60**, 925–946 (2011)
57. Cánovas, M.J., López, M.A., Parra, J., Toledo, F.J.: Calmness of the feasible set mapping for linear inequality systems. Centro de Investigación Operativa, Universidad Miguel Hernández de Elche. Preprint
58. Cánovas, M.J., Mordukhovich, B., López, M.A., Parra, J.: Variational analysis in semi-infinite and infinite programming, I: stability of linear inequality systems of feasible solutions. SIAM J. Optim. **20**, 1504–1526 (2009)
59. Charnes, A., Cooper, W.W., Kortanek, K.O.: Duality, Haar programs, and finite sequence spaces. Proc. Natl. Acad. Sci. USA **48**, 783–786 (1962)
60. Charnes, A., Cooper, W.W., Kortanek, K.O.: Duality in semi-infinite programs and some works of Haar and Carathéodory. Manag. Sci. **9**, 209–228 (1963)
61. Coelho, C.J., Galvao, R.K.H., de Araujo, M.C.U., Pimentel, M.F., da Silva, E.C.: A linear semi-infinite programming strategy for constructing optimal wavelet transforms in multivariate calibration problems. J. Chem. Inform. Comput. Sci. **43**, 928–933 (2003)
62. da Silva, A.R.: On parametric infinite optimization. Int. Ser. Numer. Math. **72**, 83–95 (1984)
63. Daniel, J.W.: Remarks on perturbations in linear inequalities. SIAM J. Numer. Anal. **12**, 770–772 (1975)
64. Daniilidis, A., Goberna, M.A., López, M.A., Lucchetti, R.: Lower semicontinuity of the solution set mapping of linear systems relative to their domains. Set-Valued Var. Anal. **21**, 67–92 (2013)
65. Daum, S., Werner, R.: A novel feasible discretization method for linear semi-infinite programming applied to basket options pricing. Optimization **60**, 1379–1398 (2011)

66. Davidson, M.R.: Stability of the extreme point set of a polyhedron. J. Optim. Theory Appl. **90**, 357–380 (1996)
67. Dentcheva, D., Ruszczyński, A.: Optimization with stochastic dominance constraints. SIAM J. Optim. **14**, 548–566 (2003)
68. Dentcheva, D., Ruszczyński, A.: Semi-infinite probabilistic optimization: first-order stochastic dominance constraint. Optimization **53**, 583–601 (2004)
69. Dentcheva, D., Ruszczyński, A.: Portfolio optimization with stochastic dominance constraints. J. Bank. Finance **30**, 433–451 (2006)
70. DeVore, R.A., Lorentz, G.G.: Constructive Approximation. Springer, Berlin (1993)
71. Dinh, N., Goberna, M.A., López, M.A.: From linear to convex systems: consistency, Farkas Lemma and applications. J. Convex Anal. **13**, 279–290 (2006)
72. Dinh, N., Goberna, M.A., López, M.A., Son, T.Q.: New Farkas-type constraint qualifications in convex infinite programming. ESAIM Control Optim. Calc. Var. **13**, 580–597 (2007)
73. Dinh, N., Goberna, M.A., López, M.A.: On the stability of the optimal value and the optimal set in optimization problems. J. Convex Anal. **19**, 927–953 (2012)
74. Dontchev, A.L., Lewis, A.S., Rockafellar, R.T.: The radius of metric regularity. Trans. Am. Math. Soc. **355**, 493–517 (2003)
75. Dontchev, A.L., Rockafellar, R.T.: Implicit Functions and Solution Mapping. A View from Variational Analysis. Springer, New York (2009)
76. Dubois, D., Kerre, E., Mesiar, R., Prade, H.: Fuzzy interval analysis. In: Dubois, D., Prade, H. (eds.) Fundamentals of Fuzzy Sets, pp. 483–581. Kluwer, Dordrecht (2000)
77. Dubois, D., Prade, H.: The mean value of a fuzzy number. Fuzzy Sets Syst. **24**, 279–300 (1988)
78. Epelman, M., Freund, R.M.: Condition number complexity of an elementary algorithm for computing a reliable solution of a conic linear system. Math. Program. A **88**, 451–485 (2000)
79. Fang, S.C., Hu, C.F., Wang, H.F., Wu, S.Y.: Linear programming with fuzzy coefficients in constraints. Comput. Math. Appl. **37**, 63–76 (1999)
80. Feyzioglu, O., Altinel, I.K., Ozekici, S.: The design of optimum component test plans for system reliability. Comput. Stat. Data Anal. **50**, 3099–3112 (2006)
81. Feyzioglu, O., Altinel, I.K., Ozekici, S.: Optimum component test plans for phased-mission systems. Eur. J. Oper. Res. **185**, 255–265 (2008)
82. Fischer, T.: Contributions to semi-infinite linear optimization. Meth. Verf. Math. Phys. **27**, 175–199 (1983)
83. Floudas, C.A., Stein, O.: The adaptive convexification algorithm: a feasible point method for semi-infinite programming. SIAM Optim. **18**, 1187–1208 (2007)
84. Freund, R.M., Vera, J.R.: Some characterizations and properties of the "distance to ill-posedness". Math. Program. A **86**, 225–260 (1999)
85. Gal, T.: Postoptimal Analyses, Parametric Programming, and Related Topics: Degeneracy, Multicriteria Decision Making, Redundancy, 2nd edn. Walter de Gruyter, New York (1995)
86. Gayá, V.E., López, M.A., Vera de Serio, V.N.: Stability in convex semi-infinite programming and rates of convergence of optimal solutions of discretized finite subproblems. Optimization **52**, 693–713 (2003)
87. Gauvin, J.: Formulae for the sensitivity analysis of linear programming problems. In: Lassonde, M. (ed.) Approximation, Optimization and Mathematical Economics, pp. 117–120. Physica-Verlag, Berlin (2001)
88. Ghaffari Hadigheh, A., Terlaky, T.: Sensitivity analysis in linear optimization: invariant support set intervals. Eur. J. Oper. Res. **169**, 1158–1175 (2006)
89. Ghaffari Hadigheh, A., Romanko, O., Terlaky, T.: Sensitivity analysis in convex quadratic optimization: simultaneous perturbation of the objective and right-hand-side vectors. Algorithmic Oper. Res. **2**, 94–111 (2007)
90. Glashoff, K., Gustafson, S.-A.: Linear Optimization and Approximation. Springer, Berlin (1983)

91. Goberna, M.A.: Linear semi-infinite optimization: recent advances. In: Rubinov, A., Jeyaku-mar, V. (eds.) Continuous Optimization: Current Trends and Applications, pp. 3–22. Springer, Berlin (2005)
92. Goberna, M.A.: Post-optimal analysis of linear semi-infinite programs. In: Chinchuluun, A., Pardalos, P.M., Enkhbat, R., Tseveendorj, I. (eds.) Optimization and Optimal Control: Theory and Applications, pp. 23–54. Springer, Berlin (2010)
93. Goberna, M.A., Gómez, S., Guerra, F., Todorov, M.I.: Sensitivity analysis in linear semi-infinite programming: perturbing cost and right-hand-side coefficients. Eur. J. Oper. Res. **181**, 1069–1085 (2007)
94. Goberna, M.A., Jeyakumar, V., Dinh, N.: Dual characterizations of set containments with strict convex inequalities. J. Global Optim. **34**, 33–54 (2006)
95. Goberna, M.A., Jeyakumar, V., Li, G., López, M.A.: Robust linear semi-infinite programming duality under uncertainty. Math. Program. B **139**, 185–203 (2013)
96. Goberna, M.A., Jeyakumar, V., Li, G.Y., Vicente-Pérez, J.: Robust solutions of uncertain multi-objective linear semi-infinite programming. School of Mathematics, UNSW, Sydney. Preprint
97. Goberna, M.A., Jeyakumar, V., Li, G.Y., Vicente-Pérez, J.: Robust solutions to multi-objective linear programs with uncertain data. School of Mathematics, UNSW, Sydney. Preprint
98. Goberna, M.A., Jornet, V.: Geometric fundamentals of the simplex method in semi-infinite programming. OR Spektrum **10**, 145–152 (1988)
99. Goberna, M.A., Larriqueta, M., Vera de Serio, V.: On the stability of the boundary of the feasible set in linear optimization. Set-Valued Anal. **11**, 203–223 (2003)
100. Goberna, M.A., Larriqueta, M., Vera de Serio, V.: On the stability of the extreme point set in linear optimization. SIAM J. Optim. **15**, 1155–1169 (2005)
101. Goberna, M.A., López, M.A.: Topological stability of linear semi-infinite inequality systems. J. Optim. Theory Appl. **89**, 227–236 (1996)
102. Goberna, M.A., López, M.A.: Linear Semi-infinite Optimization. Wiley, Chichester (1998)
103. Goberna, M.A., López, M.A., Todorov, M.I.: Stability theory for linear inequality systems. SIAM J. Matrix Anal. Appl. **17**, 730–743 (1996)
104. Goberna, M.A., López, M.A., Todorov, M.I.: Stability theory for linear inequality systems. II: upper semicontinuity of the solution set mapping. SIAM J. Optim. **7**, 1138–1151 (1997)
105. Goberna, M.A., López, M.A., Todorov, M.I.: On the stability of the feasible set in linear optimization. Set-Valued Anal. **9**, 75–99 (2001)
106. Goberna, M.A., López, M.A., Todorov, M.I.: On the stability of closed-convex-valued mappings and the associated boundaries. J. Math. Anal. Appl. **306**, 502–515 (2005)
107. Goberna, M.A., López, M.A., Volle, M.: Primal attainment in convex infinite optimization duality. J. Convex Anal. **21**, in press (2014) (unknown DOI)
108. Goberna, M.A., Martínex-Legaz, J.E., Vera de Serio, V.N.: On the Voronoi mapping. Department of Statistics and Operations Research, University of Alicante, Spain. Preprint
109. Goberna, M.A., Rodríguez, M.M.L., Vera de Serio, V.N.: Voronoi cells of arbitrary sets via linear inequality systems. Linear Algebra Appl. **436**, 2169–2186 (2012)
110. Goberna, M.A., Terlaky, T., Todorov, M.I.: Sensitivity analysis in linear semi-infinite programming via partitions. Math. Oper. Res. **35**, 14–25 (2010)
111. Goberna, M.A., Todorov M.I.: Primal, dual and primal-dual partitions in continuous linear optimization. Optimization **56**, 617–628 (2007)
112. Goberna, M.A., Todorov, M.I.: Generic primal-dual solvability in continuous linear semi-infinite programming. Optimization, **57**, 1–10 (2008)
113. Goberna, M.A., Todorov, M.I.: Primal-dual stability in continuous linear optimization. Math. Program. B **116**, 129–146 (2009)
114. Goberna, M.A., Todorov, M.I., Vera de Serio, V.N.: On the stability of the convex hull of set-valued mappings. SIAM J. Optim. **17**, 147–158 (2006)
115. Goberna, M.A., Todorov, M.I., Vera de Serio, V.N.: On stable uniqueness in linear semi-infinite optimization. J. Global Optim. **53**, 347–361 (2012)
116. Goberna, M.A., Vera de Serio, V.N.: On the stability of Voronoi cells. Top **20**, 411–425 (2012)

117. Goldfarb, D., Scheinberg, K.: On parametric semidefinite programming. Appl. Numer. Math. **29**, 361–377 (1999)
118. Gol'shtein, E.G.: Theory of Convex Programming. Translations of Mathematical Monographs, vol. 36. American Mathematical Society, Providence (1972)
119. Greenberg, H.: The use of the optimal partition in a linear programming solution for postoptimal analysis. Oper. Res. Lett. **15**, 179–185 (1994)
120. Greenberg, H.J.: Matrix sensitivity analysis from an interior solution of a linear program. INFORMS J. Comput. **11**, 316–327 (1999)
121. Greenberg, H.J.: Simultaneous primal-dual right-hand-side sensitivity analysis from a strict complementary solution of a linear program. SIAM J. Optim. **10**, 427–442 (2000)
122. Greenberg, H.J., Holder, A., Roos, C., Terlaky, T.: On the dimension of the set of rim perturbations for optimal partition invariance. SIAM J. Optim. **9**, 207–216 (1998)
123. Greenberg, H.J., Pierskalla, W.P.: Stability theory for infinitely constrained mathematical programs. J. Optim. Theory Appl. **16**, 409–428 (1975)
124. Guddat, J., Jongen, H.Th., Rückmann, J.-J.: On stability and stationary points in nonlinear optimization. J. Aust. Math. Soc. B **28**, 36–56 (1986)
125. Guo, P., Huang G.H.: Interval-parameter semi-infinite fuzzy-stochastic mixed-integer programming approach for environmental management under multiple uncertainties. Waste Manag. **30**, 521–531 (2010)
126. Guo, P., Huang, G.H., He, L.: ISMISIP: an inexact stochastic mixed integer linear semi-infinite programming approach for solid waste management and planning under uncertainty. Stoch. Environ. Res. Risk Assess. **22**, 759–775 (2008)
127. Hantoute, A., López, M.A.: Characterization of total ill-posedness in linear semi-infinite optimization. J. Comput. Appl. Math. **217**, 350–364 (2008)
128. Hantoute, A., López, M.A., Zălinescu, C.: Subdifferential calculus rules in convex analysis: a unifying approach via pointwise supremum functions. SIAM J. Optim. **19**, 863–882 (2008)
129. He, L., Huang, G.H.: Optimization of regional waste management systems based on inexact semi-infinite programming. Can. J. Civil Eng. **35**, 987–998 (2008)
130. He, L., Huang, G.H., Lu, H.: Bivariate interval semi-infinite programming with an application to environmental decision-making analysis. Eur. J. Oper. Res. **211**, 452–465 (2011)
131. Helbig, S.: Stability in disjunctive linear optimization I: continuity of the feasible set. Optimization **21**, 855–869 (1990)
132. Hettich, R., Zencke, P.: Numerische Methoden der Approximation und der Semi-Infiniten Optimierung. Teubner, Stuttgart (1982)
133. Hirabayashi, R., Jongen, H.Th., Shida, M.: Stability for linearly constrained optimization problems. Math. Program. A **66**, 351–360 (1994)
134. Hiriart-Urrity, J.-B., Lemaréchal, C.: Convex Analysis and Minimization Algorithms I, II. Springer, New York (1993)
135. Hoffman, A.J.: On approximate solutions of systems of linear inequalities. J. Res. Nat. Bur. Stand. **49**, 263–265 (1952)
136. Hogan, W.W.: The continuity of the perturbation function of a convex program. Oper. Res. **21**, 351–352 (1973)
137. Homem-de-Mello, T., Mehrotra, S.: A cutting-surface method for uncertain linear programs with polyhedral stochastic dominance constraints. SIAM J. Optim. **20**, 1250–1273 (2009)
138. Hu, C.F., Fang, S.C.: A relaxed cutting plane algorithm for solving fuzzy inequality systems. Optimization **45**, 89–106 (1999)
139. Hu, H.: Perturbation analysis of global error bounds for systems of linear inequalities. Math. Program. B **88**, 277–284 (2000)
140. Hu, H., Wang, Q.: On approximate solutions of infinite systems of linear inequalities. Linear Algebra Appl. **114/115**, 429–438 (1989)
141. Hu, J., Homem-de-Mello, T., Mehrotra, S.: Sample average approximation of stochastic dominance constrained programs. Math. Program. A **133**, 171–201 (2012)
142. Huang, G.H., He, L., Zeng, G.M., Lu, H.W.: Identification of optimal urban solid waste flow schemes under impacts of energy prices. Environ. Eng. Sci. **25**, 685–695 (2008)

143. Huy, N.Q., Yao, J.-C.: Semi-infinite optimization under convex function perturbations: Lipschitz stability. J. Optim. Theory Appl. **148**, 237–256 (2011)
144. Ioffe, A.D.: On stability estimates for the regularity of maps. In: Brezis, H., Chang, K.C., Li, S.J., Rabinowitz, P. (eds.) Topological Methods, Variational Methods, and Their Applications, pp. 133–142. World Scientific, River Edge (2003)
145. Ioffe, A.D., Lucchetti, R.: Typical convex program is very well posed. Math. Program. B **104**, 483–499 (2005)
146. Jansen, B., de Jong, J.J., Roos, C., Terlaky, T.: Sensitivity analysis in linear programming: just be careful! Eur. J. Oper. Res. **101**, 15–28 (1997)
147. Jansen, B., Roos, C., Terlaky, T.: An interior point approach to postoptimal and parametric analysis in linear programming. Technical Report, Eötvös University, Budapest, Hungary (1992)
148. Jansen, B., Roos, C., Terlaky, T., Vial, J.-Ph.: Interior-point methodology for linear programming: duality, sensitivity analysis and computational aspects. Technical Report 93-28, Delft University of Technology, Faculty of Technical Mathematics and Computer Science, Delft (1993)
149. Jaume, D., Puente, R.: Representability of convex sets by analytical linear inequality systems. Linear Algebra Appl. **380**, 135–150 (2004)
150. Jeyakumar, V., Li, G.: Strong duality in robust convex programming: complete characterizations. SIAM J. Optim. **20**, 3384–3407 (2010)
151. Jongen, H.Th., Rückmann, J.-J.: On stability and deformation in semi-infinite optimization. In: Reemtsen, R., Rückmann, J.J. (eds.) Semi-infinite Programming, pp. 29–67. Kluwer, Boston (1998)
152. Jongen, H.Th., Twilt, F., Weber, G.-H.: Semi-infinite optimization: structure and stability of the feasible set. J. Optim. Theory Appl. **72**, 529–552 (1992)
153. Jongen, H.Th., Weber, G.-H.: Nonlinear optimization: characterization of structural stability. J. Global Optim. **1**, 47–64 (1991)
154. Juárez, E.L., Todorov, M.I.: Characterization of the feasible set mapping in one class of semi-infinite optimization problems. Top **12**, 135–147 (2004)
155. Karimi, A., Galdos, G.: Fixed-order H_∞ controller design for nonparametric models by convex optimization. Automatica **46**, 1388–1394 (2010)
156. Klatte, D., Henrion, R.: Regularity and stability in nonlinear semi-infinite optimization. In: Reemtsen, R., Rückmann, J.J. (eds.) Semi-infinite Programming, pp. 69–102. Kluwer, Boston (1998)
157. Klatte, D., Kummer, B.: Stability properties of infima and optimal solutions of parametric optimization problems. In: Demyanov, V.F., Pallaschke, D. (eds.) Nondifferentiable Optimization: Motivations and Applications, pp. 215–229. Springer, Berlin (1985)
158. Klatte, D., Kummer, B.: Nonsmooth Equations in Optimization. Kluwer, Dordrecht (2002)
159. Klatte, D., Kummer, B.: Optimization methods and stability of inclusions in Banach spaces. Math. Program. B **117**, 305–330 (2009)
160. Kojima, M.: Strongly stable stationary solutions in nonlinear programs. In: Robinson, S.M. (ed.) Analysis and Computation of Fixed Points, pp. 93–138. Academic, New York (1980)
161. Kortanek, K.O.: Constructing a perfect duality in infinite programming. Appl. Math. Optim. **3**, 357–372 (1976/1977)
162. Kortanek, K.O., Medvedev, V.G.: Building and Using Dynamic Interest Rate Models. Wiley, Chichester (2001)
163. Krabs, W.: Optimization and Approximation. Wiley, New York (1979)
164. Krishnan, K., Mitchel, J.E.: A semidefinite programming based polyhedral cut and price approach for the maxcut problem. Comput. Optim. Appl. **33**, 51–71 (2006)
165. Kruger, A., Ngai, H.V., Thera, M.: Stability of error bounds for convex constraint systems in Banach spaces. SIAM J. Optim. **20**, 3280–3296 (2010)
166. Larriqueta, M., Vera de Serio, V.N.: On metric regularity and the boundary of the feasible set in linear optimization. Set-Valued Var. Anal. doi:10.1007/s11228-013-0241-8 (in press)

167. Leibfritz, F., Maruhn, J.H.: A successive SDP-NSDP approach to a robust optimization problem in finance. Comput. Optim. Appl. **44**, 443–466 (2009)
168. León, T., Liern, V., Marco, P., Segura, J.V., Vercher, E.: A downside risk approach for the portfolio selection problem with fuzzy returns. Fuzzy Econ. Rev. **9**, 61–77 (2008)
169. León, T., Sanmatías, S., Vercher, E.: On the numerical treatment of linearly constrained semi-infinite optimization problems. Eur. J. Oper. Res. **121**, 78–91 (2000)
170. León, T., Vercher, E.: Optimization under uncertainty and linear semi-infinite programming: a survey. In: Goberna, M.A., López, M.A. (eds.) Semi-infinite Programming: Recent Advances, pp. 327–348. Kluwer, Dordrecht (2001)
171. León, T., Vercher, E.: Solving a class of fuzzy linear programs by using semi-infinite programming techniques. Fuzzy Sets Syst. **6**, 235–252 (2004)
172. Levy, A.B., Poliquin, R.A.: Characterizing the single-valuedness of multifunctions. Set-Valued Anal. **5**, 351–364 (1997)
173. Li, H., Huang, G.H., Lu, H.: Bivariate interval semi-infinite programming with an application to environmental decision-making analysis. Eur. J. Oper. Res. **211**, 452–465 (2011)
174. Li, W.: The sharp Lipschitz constants for feasible and optimal solutions of a perturbed linear program. SIAM J. Optim. **187**, 15–40 (1993)
175. Li, C., Ng, K.F.: On constraint qualification for an infinite system of convex inequalities in a Banach Space. SIAM J. Optim. **15**, 488–512 (2005)
176. López, M.A.: Stability in linear optimization and related topics. A personal tour. Top **20**, 217–244 (2012)
177. López, M.A, Mira, J.A., Torregrosa, G.: On the stability of infinite-dimensional linear inequality systems. Numer. Funct. Anal. Optim. **19**, 1065–1077 (1985–1986)
178. López, M.A., Still, G.: Semi-infinite programming. Eur. J. Oper. Res. **180**, 491–518 (2007)
179. López, M.A., Vera de Serio, V.: Stability of the feasible set mapping in convex semi-infinite programming. In: Goberna, M.A., López, M.A. (eds.) Semi-infinite Programming. Recent Advances, pp. 101–120. Kluwer, Dordrecht (2001)
180. Lucchetti, R.: Convexity and Well-Posed Problems. Springer, New York (2006)
181. Lucchetti, R., Viossat, Y.: Stable correlated equilibria: the zero-sum case. Milano Politecnico, 2011. Technical Report
182. Luo, Z.-Q., Roos, C., Terlaky, T.: Complexity analysis of a logarithmic barrier decomposition method for semi-infinite linear programming. Appl. Numer. Math. **29**, 379–394 (1999)
183. Luo, Z.-Q., Tseng, P.: Perturbation analysis of a condition number for linear systems. SIAM J. Matrix Anal. Appl. **15**, 636–660 (1994)
184. Mangasarian, O.L., Wild, E.W.: Nonlinear knowledge in kernel approximation. IEEE Trans. Neural Netw. **18**, 300–306 (2007)
185. Mangasarian, O.L., Wild, E.W.: Nonlinear knowledge-based classification. IEEE Trans. Neural Netw. **19**, 1826–1832 (2008)
186. Maruhn, J.H.: Robust Static Super-Replication of Barrier Options. De Gruyter, Berlin (2009)
187. Mira, J.A., Mora, G.: Stability of linear inequality systems measured by the Hausdorff metric. Set-Valued Anal. **8**, 253–266 (2000)
188. Monteiro, R., Mehotra, S.: A generalized parametric analysis approach and its implication to sensitivity analysis in interior point methods. Math. Program. A **72**, 65–82 (1996)
189. Mordukhovich, B.S.: Coderivative analysis of variational systems. J. Global Optim. **28**, 347–362 (2004)
190. Norbedo, S., Zang, Z.Q., Claesson, I.: A semi-infinite quadratic programming algorithm with applications to array pattern synthesis. IEEE Trans. Circuits Syst. II Analog Digital Signal Process. **48**, 225–232 (2001)
191. Ochoa, P.D., Vera de Serio, V.N.: Stability of the primal-dual partition in linear semi-infinite programming. Optimization **61**, 1449–1465 (2012)
192. Oskoorouchi, M.R., Ghaffari, H.R., Terlaky, T., Aleman, D.M.: An interior point constraint generation algorithm for semi-infinite optimization with health-care application. Oper. Res. **59**, 1184–1197 (2011)

193. Ozogur, S., Weber, G.W.: On numerical optimization theory of infinite kernel learning. J. Global Optim. **48**, 215–239 (2010)
194. Ozogur, S., Weber, G.W.: Infinite kernel learning via infinite and semi-infinite programming. Optim. Methods Softw. **25**, 937–970 (2010)
195. Parks, M.L., Jr., Soyster, A.L.: Semi-infinite and fuzzy set programming. In: Fiacco, A.V., Kortanek, K.O. (eds.) Semi-infinite Programming and Applications, pp. 219–235. Springer, New York (1983)
196. Peña, J., Vera, J.C., Zuluaga, L.F.: Static-arbitrage lower bounds on the prices of basket options via linear programming. Quant. Finance **10**, 819–827 (2010)
197. Powell, M.J.D.: Approximation Theory and Methods. Cambridge University Press, Cambridge (1981)
198. Puente, R.: Cyclic convex bodies and optimization moment problems. Linear Algebra Appl. **426**, 596–609 (2007)
199. Puente, R., Vera de Serio, V.N.: Locally Farkas-Minkowski linear inequality systems. Top **7**, 103–121 (1999)
200. Renegar, J.: Some perturbation theory for linear programming. Math. Program. A **65**, 73–91 (1994)
201. Renegar, J.: Linear programming, complexity theory and elementary functional analysis. Math. Program. A **70**, 279–351 (1995)
202. Robinson, S.M.: Stability theory for systems of inequalities. Part I: linear systems. SIAM J. Numer. Anal. **12**, 754–769 (1975)
203. Robinson, S.M.: Stability theory for systems of inequalities. Part II: differentiable nonlinear systems. SIAM J. Numer. Anal. **13**, 497–513 (1976)
204. Robinson, S.M.: Some continuity properties of polyhedral multifunctions. Math. Program. Study **14**, 206–214 (1981)
205. Rockafellar, R.T.: Convex Analysis. Princeton University Press, Princeton (1970)
206. Rockafellar, R.T., Uryasev, S.: Optimization of conditional value-at-risk. J. Risk **2**, 21–41 (2000)
207. Rockafellar, R.T., Wets, R.J.-B.: Variational Analysis. Springer, Berlin (1998)
208. Roos, C., Terlaky, T., Vial, J.-Ph.: Theory and Algorithms for Linear optimization: An Interior Point Approach. Wiley, Chichester (1997)
209. Rubinstein, G.S.: A comment on Voigt's paper "a duality theorem for linear semi-infinite programming". Optimization **12**, 31–32 (1981)
210. Shapiro, A.: Directional differentiability of the optimal value function in convex semi-infinite programming. Math. Program. A **70**, 149–157 (1995)
211. Shapiro, A., Dentcheva, D., Ruszczyński, A.: Lectures on Stochastic Programming: Modeling and Theory. MOS-SIAM Series on Optimization. SIAM, Philadelphia (2009)
212. Sharkey, T.C.: Infinite linear programs. In: Cochran, J.J. (ed.) Wiley Encyclopedia of Operations Research and Management Science, pp. 1–11. Wiley, New York (2010)
213. Sonnenburg, S., Rätsch, G., Schäfer, C., Schölkopf, B.: Large scale multiple kernel learning. J. Mach. Learn. Res. **7**, 1531–1565 (2006)
214. Soyster, A.L.: Convex programming with set-inclusive constraints and applications to inexact linear programming. Oper. Res. **21**, 1154–1157 (1973)
215. Stein, O., Still, G.: Solving semi-infinite optimization problems with interior point techniques. SIAM J. Control Optim. **42**, 769–788 (2003)
216. Tanaka, H., Okuda, T., Asai, K.: On fuzzy mathematical programming. J. Cybern. **3**, 37–46 (1974)
217. Tian, Y., Shi, Y., Liu, X.: Recent advances on support vector machines research. Technol. Econ. Dev. Econ. **18**, 5–33 (2012)
218. Tichatschke, R.: Lineare Semi-Infinite Optimierungsaufgaben und ihre Anwendungen in der Approximationstheorie. Wissenschaftliche Schriftenreihe der Technischen Hochschule, Karl-Marx-Stadt (1981)
219. Tichatschke, R., Hettich, R., Still, G.: Connections between generalized, inexact and semi-infinite linear programming. Math. Methods Oper. Res. **33**, 367–382 (1989)

220. Todorov, M.I.: Generic existence and uniqueness of the solution set to linear semi-infinite optimization problems. Numer. Funct. Anal. Optim. **8**, 27–39 (1985/1986)

221. Toledo, F.J.: Some results on Lipschitz properties of the optimal values in semi-infinite programming. Optim. Methods Softw. **23**, 811–820 (2008)

222. Tuy, H.: Stability property of a system of inequalities. Math. Oper. Stat. Ser. Opt. **8**, 27–39 (1977)

223. Vaz, A., Fernandes, E., Gomes, M.: SIPAMPL: semi-infinite programming with AMPL. ACM Trans. Math. Softw. **30**, 47–61 (2004)

224. Vercher, E.: Portfolios with fuzzy returns: selection strategies based on semi-infinite programming. J. Comput. Appl. Math. **217**, 381–393 (2008)

225. Vercher, E., Bermúdez, J.D.: Fuzzy portfolio selection models: a numerical study. In: Doumpos, M., Zopounidis, C., Pardalos, P.M. (eds.) Financial Decision Making Using Computational Intelligence, pp. 245–272. Springer, New York (2012)

226. Voigt, I., Weis, S.: Polyhedral Voronoi cells. Contr. Beiträge Algebra Geom. **51**, 587–598 (2010)

227. Wan, Z., Meng, F.-Z., Hao, A.-Y., Wang, Y.-L.: Optimization of the mixture design for alumina sintering with fuzzy ingredients. Hunan Daxue Xuebao/J. Hunan Univ. Nat. Sci. **36**, 55–58 (2009)

228. Wang, X., Kerre, E.E.: Reasonable properties for the ordering of fuzzy quantities (I) & (II). Fuzzy Sets Syst. **118**, 375–385, 387–405 (2001)

229. Wu, D., Han, J.-Y., Zhu, J.-H.: Robust solutions to uncertain linear complementarity problems. Acta Math. Appl. Sin., Engl. Ser. **27**, 339–352 (2011)

230. Yildirim, E.A.: Unifying optimal partition approach to sensitivity analysis in conic optimization. J. Optim. Theory Appl. **122**, 405–423 (2004)

231. Yiu, K.F., Xioaqi, Y., Nordholm, S., Teo, K.L.: Near-field broadband beamformer design via multidimensional semi-infinite linear programming techniques. IEEE Trans. Speech Audio Process. **11**, 725–732 (2003)

232. Zălinescu, C.: On the differentiability of the support function. J. Global Optim. **57**, 719–731 (2013)

233. Zălinescu, C.: Relations between the convexity of a set and the differentiability of its support function. arXiv:1301.0810 [math.FA] (2013). http://arxiv.org/abs/1301.0810

234. Zencke, P., Hettich, R.: Directional derivatives for the value-function in semi-infinite programming. Math. Program. A **38**, 323–340 (1987)

235. Zimmermann, H.J.: Description and optimization of fuzzy systems. Int. J. Gen. Syst. **2**, 209–215 (1976)

236. Zheng, X.Y., Ng, K.F.: Metric regularity and constraint qualifications for convex inequality on Banach spaces. SIAM J. Optim. **14**, 757–772 (2003)

237. Zhu, Y., Huang, G.H., Li, Y.P., He, L., Zhang, X.X.: An interval full-infinite mixed-integer programming method for planning municipal energy systems—a case study of Beijing. Appl. Energy **88**, 2846–2862 (2011)

238. Zopounidis, C., Doumpos, M.: Multicriteria decision systems for financial problems. Top **21**, 241–261 (2013)

Index

M.A. Goberna and M.A. López, *Post-Optimal Analysis in Linear Semi-Infinite Optimization*, SpringerBriefs in Optimization, DOI 10.1007/978-1-4899-8044-1, © Miguel A. Goberna, Marco A. López 2014